D0267562

ACC. No: 05152593

It's Never
Too Late
STORIES OF PEOPLE WHO
CHANGED THE WORLD
IN LATER LIFE

It's Never Too Late

Stories of People Who Changed the World in Later Life

Chester Morganfield

FOREWORD BY
Michael Whitehall

CORONET

First published in Great Britain in 2019 by Coronet
An Imprint of Hodder & Stoughton
An Hachette UK company

1

A CIP catalogue record for this title is available from the British Library

Hardback ISBN 978 1 529 31600 1
Trade Paperback ISBN 978 1 529 31601 8
eBook ISBN 978 1 529 31603 2

Typeset in Sabon MT by Hewer Text UK Ltd, Edinburgh
Printed and bound in Great Britain by Clays Ltd, Elcograf S.p.A.

Hodder & Stoughton policy is to use papers that are natural, renewable and recyclable products and made from wood grown in sustainable forests. The logging and manufacturing processes are expected to conform to the environmental regulations of the country of origin.

Hodder & Stoughton Ltd
Carmelite House
50 Victoria Embankment
London EC4Y 0DZ

www.hodder.co.uk

Contents

Foreword

Without wishing to gush with faux modesty all over the opening pages of this intriguing and entertaining book, I was very surprised to be asked to write a foreword to it, celebrating the impressive, inventive and astonishing things people have achieved in older age. But having pondered on its subject matter, I surmised that due to being a perennial late starter – late to puberty, late to marriage, late to fatherhood, very late to a second career – I was actually quite a good fit for the job.

There is a satisfying truth to the saying, 'Good things come to those that wait.'

Late fatherhood brought me some unexpected treasures, not least the joy of reading my children bedtime stories. How else would I have discovered the wonderful Judith Kerr, (featured later) and her all-time classic, written when she was in middle age, *The Tiger Who Came to Tea*, which at one point in my life I could have recited without the aid of the text, such was its popularity in the Whitehall household.

When my son Jack asked me to appear with him at the Edinburgh Fringe Festival, I was in my early seventies and heading for a retirement of gazing out of the window wondering who I was, where I was, and why I was looking out of the window. (I also have a son Barney, who many think is just as funny as Jack, and a daughter Molly, who is on a par with her saintly mother Hilary, whom I met when I was well into my forties and still takes my breath away every time I see her.)

Jack was headlining his first big gig at the 1,200-seater EICC and thought it would be fun to fill in his afternoons doing a show together, in the 'Nicholas Parsons slot' at the Pleasance Cabaret Room.

The plan was for Jack to host a live chat show and for me to be a silent participant sitting on the stage in a comfortable Victorian armchair, similar to Madge in Barry Humphries's *The Dame Edna Everage Show*. I was doubtful. I had spent my life working first as a theatrical agent and then a TV and theatre producer and had never harboured any aspirations to appear on the other side of the curtain. However, Jack managed to talk me round; and with my reluctant agreement, *Backchat* was born.

Jack soon realised that I was never going to be a mute partner in this enterprise and the genie was quickly out of the bottle. I cultivated an exaggerated version of my grumpy and opinionated self, although Jack, and indeed the rest of my family and friends, thought that any exaggeration was difficult to detect. This live stage show then found its way onto the BBC and, in turn, this expanded into *Travels with My Father* for Netflix, which has just completed its third series.

This trajectory shares similarities with one of the subjects in this book and indeed one of our guests on *Backchat*: Baroness Trumpington. She forged an even later career as a television personality, impressively well into her eighties, having 'gone viral' after flicking Lord King the v-sign during a House of Lords debate. Ironic that this is what gained her national fame, rather than her role in the code-breaking team at Bletchley Park during the Second World War, as well as a long career in politics awash with good works and worthy causes.

Getting on television would have made my mother so proud.

'You'd make a lovely actor,' she said to me in my early twenties. She had said similar things to me about a multitude of other jobs, however, which I attempted with spectacular lack of success.

'You'd make a wonderful barrister,' she told me, as she pressed a newly bought wig and gown into my hands. I bunked out after a year, before the first set of exams. I still have the wig and gown, which are now in my children's seldom-used, dressing-up box.

Another proponent of the pot-luck approach to life was Colonel Sanders. He took an even longer route than me to find his niche and ran through a number of diverse jobs until, in his mid-sixties, he managed a gas station and came up with the idea of offering fried chicken to his customers. By the age of seventy-three, he had six hundred franchises. Despite my protestations to my children that his product was not the healthiest option available, my argument would not have been helped had they known what I have learnt from his section in this book, which is that he spent the last thirty years of his life happily eating his chicken and died aged ninety.

With my new-found infamy, I was faced with the prospect of going from one embarrassing reality TV show to another and fading from sight, as swiftly as I had arrived. Although with my advancing years, I had the perfect excuse to dodge several bullets during this period, one of which glorified in the name of *Celebrity Car Boot Sale*. Others included *Celebrity X Factor*, *Route 66 Year Olds*, and *Pointless Celebrities*, the clue being in the title. I also turned down the first series of *The Real Marigold Hotel*. Curry and I have never been friends. I commend Harry Redknapp and Stanley Johnson's resilience at surviving in the jungle, but this was certainly not for me.

I love hearing about people achieving things later in life, it gives hope to everyone. Never give up, because who knows what might be lurking around the corner? I'm a good example of that, as are all the people you are soon to read about. The power of perseverance. There is a great truth in the old adage, 'Never say never.'

The advantage of forging a new career in your 'third age', is that you require all the attributes that are the 'tool kit' of youth; flexibility, dexterity and agility, both mentally and physically. Although these may have ebbed over the passage of time, the attributes acquired by spending decades at the rockface, gaining experience, wisdom and knowledge, far outweigh these deficiencies.

Ronald Reagan was a particularly intriguing character in this very respect. As you will discover, he started out as a movie star (albeit in some modest movies), then became head of the Screen Writers Guild, Governor of California, and President of the USA at the age of seventy. He was nearly eighty by the time he left office. He is referred to as being 'the right kind of old. Genial, avuncular and steady.' I particularly enjoyed his remark when standing for President, when he assured voters that he would 'not make age an issue of this campaign', by exploiting his political opponent's 'youth and inexperience'.

Indeed, it's interesting that many of the most successful civilisations and communities over the ages have been the ones that revere and respect the standing and teachings of the elders amongst them.

It's hard to make a comeback when you haven't been anywhere and the one massive advantage about experiencing success and prominence later in life, is that you will never, ever be a has-been.

Michael Whitehall
September 2019

It's Never Too Late

Stories of People Who Changed the World in Later Life

William Ivy Baldwin

In 1928, an organisation was founded to honour the first pioneers of flight – many of whom, at that point, were still alive.

The Early Birds of Aviation required its members to have flown solo before 17 December 1916. (After which date, pilots were trained in much larger numbers for service in the First World War.)

The list of Early Birds includes some extraordinary people, reflecting the fact that in its earliest days, aviation was the pursuit of eccentrics and daredevils. None was more eccentric or more daring than William Ivy Baldwin.

Baldwin was born, simply, William Ivy in Texas in 1866. It's hard to know what compelled him to do it (details of his childhood are thin), but aged eleven or twelve, William ran away from his parents and joined a travelling circus.

The circus featured a popular balloon act: the balloonist would ascend to a dizzying height in a perfectly round gas balloon and then – with the excitable crowd below roaring with fear and suspense – leap out, and parachute back to earth.

One night, the balloonist was too drunk to perform. So the thirteen-year-old William went to the circus manager's tent and volunteered to pull off the stunt himself. The circus manager – weighing the boy's safety against the prospect of a mutinous audience – happily agreed.

William's short stature and light frame made him an ideal balloonist. He performed perfectly that night, replaced the drunk balloonist permanently, and never looked back.

William's aerial stunts were wildly popular and in time he went into business with another balloon pioneer, Scott Baldwin, and changed his own name to William Ivy Baldwin – so that together they could present themselves as 'The Baldwin Brothers'.

In the 1890s, he was persuaded to join the US military Signals Corps, deploying his balloon for reconnaissance and observation. He flew it over Cuba in 1898, when the United States was at war with Spain.

The Baldwin Brothers continued to perform. On one occasion they entertained the Emperor of Japan, who presented William with a silk kimono embroidered with balloons and parachutes.

By now William looked every inch the diminutive but dashing daredevil that he had become: he had a glint in his eye and a moustache far wider than his head. He had made several thousand parachute jumps and who knows how many balloon flights.

He needed something else. In 1903, the Wright Brothers had achieved powered flight in a machine heavier than air – albeit for a matter of seconds. Baldwin resolved to follow them.

He built his own aircraft in 1910 and in the same year achieved short flights in several states including Colorado and Nevada, where he is enshrined in the Nevada Aerospace Hall of Fame as the first person ever to fly a powered aircraft there.

(He also set a new altitude record in Nevada, although it helped that he was already 4,675 feet above sea level when he stepped into the plane.)

None of these achievements, however, qualify Baldwin for inclusion in this book. He was, after all, making a name for himself in his chosen field of leaping out of things in his early teens.

Alongside the ballooning, flying, parachuting and moustache-growing, Baldwin developed a sideline in the gentler art of tightrope-walking. He was, by all accounts, pretty good at it; no doubt this occupation, too, suited his tiny, wiry frame.

And so we come to the feat for which William Ivy Baldwin is remembered. One summer day, Baldwin installed a wire spanning South Boulder Creek in Colorado, a distance of 635 feet.

Watched by, among others, his daughter, who had insisted he lower the wire a little, he stepped out over the 125 foot chasm in ordinary clothes and plain cloth shoes. Slowly, carefully, he made his way safely to the other side. The year was 1948 and he was eighty-two years old.

Sister Wendy Beckett

In 1991, Sister Wendy Beckett lived in solitude in a caravan in the grounds of a Carmelite monastery in Norfolk. When she wasn't praying (she put in about seven hours a day) she was translating Latin. She had lived this way since 1970. She was sixty-one years old.

Despite her austere lifestyle, however, she also found time to write. Her subject: art. Her 1988 book *Contemporary Women Artists*, written with the blessing of the Church, had been very well received. That's why she was contacted in 1991 by a BBC producer working on a documentary about the National Gallery in London.

Sister Wendy appeared in that programme for about four minutes. But those four minutes were enough. A *bona fide* TV natural had stepped out of a caravan – dressed in her nun's habit – and straight into TV stardom.

The BBC immediately offered Sister Wendy her own series and soon this extraordinarily sharp and charming nun was as familiar to British audiences (and later American ones, too) as anyone else on TV in the 1990s.

'Viewers were transfixed,' wrote the *Washington Post*, by this 'bucktoothed nun in full religious habit and oversized glasses, speaking with wit, warmth – and a slight speech impediment.'

Her success could in part be explained by her exuberant authenticity and her ability to deliver unscripted art exposition, which was as profound as it was easy to understand. Her producers called her 'One Take Wendy'.

No doubt many were drawn to her by sheer curiosity. How would she describe lascivious imagery? ('Her pubic hair is so soft and fluffy.') How would she reconcile TV stardom with the hermitic life? (She simply retreated to her caravan and seclusion between productions.)

But her success could also be explained by her intelligence. Sister Wendy knew her stuff. Some established art critics of the time peered down their noses at this upstart nun, describing her work as simplistic or amateurish.

Perhaps they resented her ability to connect with audiences that might otherwise have been their own. Perhaps there was some misogyny, or the belief that an ageing nun – a consecrated virgin, no less – couldn't possibly be expected to understand art in all its worldly complexity.

They were wrong. Sister Wendy was every bit their intellectual equal. She had studied English Literature at Oxford (where she was already a nun, following a vow of silence) and graduated with a 'congratulatory first'. Her tutors, including a certain J.R.R. Tolkien, applauded as they made the award.

Her own taste in art might have been sneered at by her detractors, but to the public it was perfectly sound. She described Diego Velázquez as 'the greatest painter the world has ever seen' and Damien Hirst as 'really pathetic'. Who could argue with that?

She had also seen more of the world than many supposed. Sister Wendy taught in schools in South Africa – the country of her birth – for nearly twenty years before moving to her caravan and solitude.

In 1996, she made *Sister Wendy's Story of Painting*, a ten-part series that involved travel to at least as many countries. Her agent insisted on inserting a clause into her contract giving her time to go to mass every day. It was an ambitious series, which aired in the UK and the US and cemented Sister Wendy's stature as the world's most recognisable commentator on art.

This was a wildly improbable position for an old nun to find herself in. But Sister Wendy seems never to have been particularly surprised. Perhaps she saw the whole astonishing adventure as part of a much bigger plan?

She told BBC Radio 4's *Desert Island Discs*: 'Making television isn't the kind of prayer I would have chosen, but it's what God chose for me.' Sister Wendy Beckett died in 2018 at the age of eighty-eight.

Ray Kroc

There are at least two versions of the Ray Kroc story: the official McDonald's biography, and the rest. But what they have in common is that Kroc came late to the business that would make him not only very wealthy, but an enduring inspiration to a particular kind of entrepreneur.

'I was fifty-two years old,' he wrote in his 1977 autobiography, *Grinding It Out*. 'I had diabetes and incipient arthritis. I had lost my gall bladder and most of my thyroid gland in earlier campaigns. But I was convinced that the best was ahead of me.'

He was absolutely right. Kroc was born in 1902 in Oak Park, Illinois, to Czech immigrants. For the next fifty years he turned his hand to many things: wartime ambulance driver, pianist, paper cup salesman (for seventeen years!) and, his penultimate gig, travelling salesman for Multimix milkshake machines.

It was the milkshake machines that led him to McDonald's – a restaurant in San Bernardino, California, which had eight of them. Kroc found two brothers, Dick and Mac McDonald, and a kitchen system that blew his mind.

The McDonald brothers had devised a short menu and ultra-efficient food preparation method, which allowed them to turn out hamburgers and fries quickly and remarkably consistently.

'I was fascinated by the simplicity and effectiveness of the system they described that night,' he wrote. 'Each step in producing the limited menu was stripped down to its essence and accomplished with a minimum of effort.'

In the official McDonald's story, the brothers wanted Kroc's assistance in finding potential franchise partners. So Kroc – happy to help – founded McDonald's System, Inc. in 1955 and went to work. He soon bought them out and the rest is fast-food history.

Other versions of the story, including one told in the 2016 biopic *The Founder*, suggest Kroc was far more ruthless; badgering the brothers into cutting him in on the business and then edging them out of it for a paltry $2.7 million. That was a lot of money back then, of course, but nothing compared to what was to come.

Kroc might not have been a nice guy. He is sometimes described as 'mercurial'. He was clearly driven. And ultimately, whether flooding the world with cheap burgers has made it a better place is definitely questionable.

But he was a brilliant businessman, there's no doubt about

it. His model of franchising allowed McDonald's to spread everywhere, quickly. He obsessed over every corner of his empire, streamlining and perfecting the operation.

In 1998, *Time* magazine included Kroc in a list of the 100 most influential people of the twentieth century. 'He helped change American business and eating habits,' said the *New York Times*, 'by deftly orchestrating the purveying of billions of small beef patties.'

Aspiring business owners admired Kroc's persistence, his attention to detail and – above all – his mantra that hard work and grit can take you anywhere. 'Nothing in the world can take the place of persistence,' he said. 'Talent will not; nothing is more common than unsuccessful men with talent.' He had a point.

Kroc is also considered the stand-out shining example of someone who proves the idea sitting behind this book; that it's never too late to get started. In truth, of course, although he did meet his destiny in late middle-age, his experience up to that point had put him in a strong position to capitalise on an opportunity when one finally came his way.

'People have marvelled at the fact that I didn't start McDonald's until I was fifty-two years old,' he wrote, 'and then I became a success overnight.

'But I was just like a lot of show business personalities who work away quietly at their craft for years, and then, suddenly, they get the right break and make it big. I was an overnight success all right, but thirty years is a long, long night.'

Andrea Camilleri

As the *Guardian* noted in its obituary of Andrea Camilleri, had he died in his fifties, his passing would not have made news.

But when he died in 2019 at the age of ninety-three, obituaries were published around the world. A lot had happened in the final third of Andrea Camilleri's life.

Andrea Camilleri was born in 1925 in the Sicilian coastal town of Porto Empedocle. His father, Giuseppe, worked for the coastguard. He was also a committed fascist and admirer of Mussolini – at least, he was until 1938.

That year, the young Andrea relayed to his father how a schoolfriend had been barred from studying because of his Jewish roots. 'My father hit the roof,' Camilleri recalled, 'saying, "That bastard," referring to Mussolini. "The Jews are just like us!"' Camilleri readily agreed, expressing his own distaste for authority by throwing an egg at the school crucifix.

After the war, in which he was nearly killed by Allied bombs during the liberation of Sicily, Camilleri embraced first communism (he would be a life-long anti-fascist) and then theatre, studying stage and film direction in Rome.

Camilleri would spend most of his working life as a drama teacher and director at the prestigious Accademia Nazionale d'Arte Drammatica in Rome. He was an influential figure in Italian television and theatre; hardly unsuccessful. But he aspired to a literary career, and here he struggled.

His first book was a complex historical novel written in difficult Sicilian dialect. Even with a glossary, which his publisher insisted on including, the book was unintelligible to most Italians and, reasonably enough, widely ignored.

Unwilling to give up on the idea, however, Camilleri published a second historical novel that was no less complex – nor any more readable – than the first. Unsurprisingly, perhaps, both books sunk without trace.

But then, in 1994, as he was approaching seventy, his theatre and teaching career behind him, Camilleri decided to set himself some homework: he would, he resolved, try his hand at writing an out-and-out thriller. He would, he said, apply 'perfect logic' to the plot.

The result was *The Shape of Water*, a murder mystery featuring, for the first time, Inspector Montalbano.

Inspector Montalbano is a fractious Sicilian detective with a love of fine food, which he prefers to eat in silence. He serves in Vigàta, an imaginary Sicilian town not unlike Camilleri's Porto Empedocle.

Montalbano was Camilleri's first novel attempt in over a decade and he must have expected it to meet the same fate as the others: immediate obscurity. Instead, of course, he had a hit on his hands.

Camilleri might have stopped here. He'd proved his point. And he was an old man. Instead, magnificently, he upped the tempo, almost as if he was making up for all those many decades now lost to him as a writer.

Between 1994 and his death in 2019, he wrote thirty books in the Inspector Montalbano series and many others, too. There were periods in his eighties, it's said, when he was writing eight books a year.

The Inspector Montalbano series was adapted for Italian television in 2000, after which sales soared even higher. Camilleri has sold many millions of books worldwide, bringing him, in old age, not just fame but also fortune.

He consistently dedicated the success of the fictional detective to his late father, whom, he says, Montalbano is modelled on. Camilleri is surely himself an inspiration to any aspiring writer over seventy.

Barbara Beskind

The world is designed for middle-aged men. That's because most things – from trains to chairs, from phones to forks – were designed by middle-aged men.

The world is not, for the most part, designed for women or the disabled or children – or the elderly. Barbara Beskind is working on that problem. Beskind has a pretty good idea what the elderly need. She's ninety-five years old.

Beskind was born in 1924 in Washington DC. Her childhood coincided with the Depression and after her father lost his job at the FBI, she felt it.

She told the BBC's *Outlook* programme: 'We had no money to buy anything, so we were all problem-solvers right from the beginning. There was no way around it.'

It's striking how many late achievers of a certain age attribute their life-long ingenuity to early poverty. In Beskind's case, there's a very clear link, and she resolved early on to become a great inventor.

But this was the 1930s and, back then, the idea that a girl would become an inventor was considered laughable. So instead, she studied home economics at Syracuse University after which, in 1945, she got the break she needed. She joined the United States Army.

The Second World War had made invalids of a great number of men. Many were treated at the famous Walter Reed Army Medical Center, where Beskind was deployed to work in occupational therapy.

It was an inspired assignment for this inventive but caring woman, who immediately set about designing equipment that could help men to function in a world for which they were no longer built.

Beskind left the Army as a major in 1966 at the age of forty-two. This was to be the first of several retirements. She married at fifty-two and set herself up in private practice, eventually moving to New Jersey. She indulged several other passions, from writing to art, and didn't ever stop working.

Then, in 2013, when she was almost ninety years old, she watched a TV show that described the work of a Silicon Valley global design firm called IDEO. She immediately wrote to the founder and suggested, with charming politeness, that perhaps they might consider taking her on. And they did.

Beskind works on improving products for the elderly. It is important work. She collaborates with designers sixty or seventy years her junior, offering insights that come both from her long experience, and her long life.

As she told *Outlook*: 'No one can expect, at a young age, to put themselves in the shoes of an elderly person and sense what it's like.' Designing products for the people who will actually use them – rather than for a generic adult male human – isn't simply a matter of comfort; it can be life-saving.

Beskind is an inspiration for continuing to work, and for having the gumption to demand that work even as she approached her nineties. But she is also an inspiration because of the nature of that work – making products that suit the elderly – and the message it sends: that older people, even the very old, have a great deal to contribute.

Ian McKellen

Ian McKellen found his passion, and his success, early. After growing up in Bolton in the 1940s, he won a scholarship to Cambridge, where he immediately shone as an actor.

He rejected drama school – he didn't need it – and went directly to the stage and theatre stardom: the RSC, the National, the West End, Broadway, a CBE and a knighthood.

But the kind of film career that brought far greater fame to many of his contemporaries, such as Peter O'Toole and Albert Finney, eluded him.

He could be forgiven for claiming he rejected Hollywood, preferring the seriousness of theatre, but he is more honest. 'It's my impression that I've done every job I've been asked to do,' he says.

Then in 1988, when he was almost fifty, something changed. In a radio debate about homophobia, he declared, 'Let's not talk in the abstract,' and outed himself.

And so, in middle-age, he embarked on an energetic second career as an activist. He helped found Stonewall, which campaigns for the equality of LGBT people. He tore Leviticus from the Bible during a one-man show. He made speeches, took part in pride marches, lobbied politicians and wasted few opportunities to use his celebrity to push his message.

'Feeling that it's appropriate to disguise something so central to your nature means that you yourself are homophobic,' he has said. 'That you don't like yourself.'

He was no longer hiding, off stage or on it. Some detected an even deeper humanity in his performances. And he began to think differently about his craft.

His style – reaching all the way to the back of the theatre – didn't work for the camera. He worked on it, taking small parts in big movies to see how it should be done.

Then, in 1995, he made *Richard III*, his Hollywood breakthrough. He was fifty-six. In 1999, he was offered Magneto in the *X-Men* series – and then, of course, Gandalf in Peter Jackson's epic *Lord of the Rings* films. Ian McKellen was now a fully-fledged global celebrity.

Not that it appears to have been his main aim. McKellen was late to film success and late to full-throated activism. Perhaps he couldn't have achieved one without the other. His main aim now, it seems, is simply to keep going.

'Acting is more important to me than ever,' he told the *Daily Telegraph* in 2013. 'If I weren't doing it, then I really would be on the slippery slope . . . the end of which is The End. So it's a matter of life and death, a bit.'

McKellen toured eighty UK theatres in 2018, his eightieth year, with a one-man show – in part, he said, to repay a debt to regional theatre. But was it also a farewell tour?

'Absolutely not,' he said. 'It's more like "oh, hello again".'

Gladys Burrill

Gladys Burrill was born in the United States, to Finnish immigrants, in 1918. She didn't have an easy time of it. Her father had black lung from working in the mines. Gladys suffered from polio. Her childhood was spent helping her parents scratch a living from a dusty farm.

'So much of my life was spent in poverty,' she told the *Oregonian*, 'it's just the way it was.' But what she did have – throughout her whole life – was access to wild places and nature, and an appetite to explore it.

She married and had five children, but she always found time to run, hike and climb regularly. Together with her husband, Eugene, she travelled, falling in love with Hawaii (where, in later life, she would spend a good deal of time).

Her children eventually provided Gladys with eighteen grandchildren – and they in turn produced dozens of great-grandchildren.

And then, in her eighties, Gladys decided to run a marathon. She chose Honolulu and covered the 26.2 miles in 2004 at the age of eighty-six. Then, the next year, she did it again. And again. 'I just got hooked on it and kept going,' she said.

Gladys Burrill ran seven marathons in all including, in 2010, completing the Honolulu marathon in 9 hours, 53 minutes and 16 seconds. By then she was ninety-two years old.

A representative of Guinness World Records was on hand to confirm that Gladys, who by now was known to many as 'The Gladyator', had set a new world record; the oldest woman ever to complete a marathon.

She put her longevity and late-life athletic success down to her state of mind. 'It's so important to think positive,' she said. 'It's easy to get discouraged and be negative. It makes such a difference in how you feel and your outlook on everything.' Her advice to other elderly people was simple: 'Just get out there and walk or run.'

Burrill didn't trust positive thinking alone to keep her in good health. She trained regularly well into her nineties, power-walking up to fifty miles a week, and adopted a mostly vegetarian diet. She never drank alcohol.

Now over 100, the Gladyator has had to hang up her running shoes. But she remains in good health and, it seems, in good spirits. Her official record still stands, providing inspiration to anyone under ninety-two wondering whether they're too old to attempt a marathon. 'Age is only a number,' she has said, 'people can be old at forty.'

There have been other aged marathon runners. Fauja Singh is a British Sikh man, who completed the Toronto Waterfront marathon in 2011 at the of age of 100. Guinness World Records have not recognised his achievement, however, because Singh was unable to produce a birth certificate proving his age. (He says birth records weren't kept in India when he was born.)

Harriette Thompson was another American woman who ran into her nineties, and may even have beaten Gladys Burrill's record (though Burrill is still recognised as the record-holder by Guinness World Records). Thompson took up marathon running in her seventies and persevered even after surviving cancer.

What all these remarkable individuals have in common is a desire to treat age as an inconvenience to be ignored.

Alexander Fleming

In 1906, the twenty-five-year-old Alexander Fleming qualified with distinction from St Mary's teaching hospital in London.

His journey to that point had not been entirely straightforward. His father – a Scottish farmer – had died when he was seven. He had followed an older brother to London at the age of thirteen, spent four years working as a shipping clerk and had even served, briefly, in the army.

But now he was a doctor. Fleming joined the staff of the St Mary's inoculation department, working under Sir Almroth Wright (a leading expert in vaccine therapy), and began to study bacteria and infection.

When Europe descended into war in 1914, Fleming was posted to British army field hospitals near the front line in France. Mechanised warfare meant vast numbers of young men were dying not from the initial impact of their wounds, but from subsequent infection.

It was a bleak time. Fleming witnessed death on an almost unimaginable scale. But he came to understand bacteria and the body's reaction to it better than anyone.

He realised that topical antiseptic medicine – very widely used at the time – often made infection worse. His attempts to get the medical profession to take his results seriously failed. Untold numbers died unnecessarily as a result.

After the war, Fleming went back to St Mary's and continued his research. For twenty years. And then, in August 1928, at the age of forty-seven, he decided to go on a two-week holiday. It is a very good thing that he did.

When he returned to his lab, he found he had left out a petri dish containing a sample of staphylococcus bacteria. While he was away, penicillium mould spores had somehow found their way into the petri dish – perhaps blowing in through an open window.

The summer temperature of the unoccupied laboratory had been almost perfect for the cultivation of both the bacteria and the mold. Both had grown. But Fleming, famously muttering, 'That's funny,' noticed that the growth of the bacteria seemed to have been inhibited by – perhaps even damaged by – the penicillium spores.

At an age that, while not exactly ancient, was at least late middle-age in the 1920s, and after twenty years of trial and error, Fleming had accidentally made a discovery which would literally save the lives of countless millions in the decades that followed.

Fleming himself did not immediately grasp the full significance of his findings. He called the precious spores 'mould juice' and set to work trying to understand how this juice might be used.

He eventually gave the juice a better name – penicillin – but for some time he saw it only as a potential topical treatment. His progress was slowed by the fact it was very difficult to produce penicillin in useable quantities.

Like his First World War observation that the widespread use of antiseptic medicine made soldiers' infections worse, not better, Fleming's discovery of penicillin was met with a collective shrug.

But when Europe descended into war again in 1939 and, again, young men were dying of infection after surviving battlefield trauma, work on penicillin accelerated and, by 1944, two other researchers – Howard Florey and Ernst Chain – had successfully turned Fleming's 'mould juice' into an effective antibiotic that could be mass produced. By this time, Alexander Fleming was in his sixties.

The consequences of Fleming's discovery of penicillin, together with Florey and Chain's development of it, are impossible to quantify. Everyday bacterial infections – which only years earlier would have led to death – could now be conquered almost overnight. In 1945, Alexander Fleming, Florey and Chain shared the Nobel Prize in medicine.

In the same year Fleming gave an interview to the *New York Times* in which he warned that the over-prescription and misuse of antibiotics could lead to bacteria becoming resistant to them.

Fleming was warning, at the very dawn of the antibacterial era, that without care that era would come to an abrupt end. He was, as he had been twice before, largely ignored. But as we now know – just like before – he was absolutely right.

The World Health Organization says: 'some of the world's most common – and potentially most dangerous – infections are proving drug-resistant.' The UN has warned that antibacterial resistance is the cause of 700,000 deaths a year and that number will 'skyrocket' to 10 million or more deaths per year within thirty years.

'We are,' said the director-general of the WHO, in 2019, 'at a critical point in the fight to protect some of our most essential medicines.' Alexander Fleming was a man worth listening to.

Whina Cooper

On 14 September 1975, fifty Maori protestors set off from Te Hāpua in the far north of New Zealand. They were marching on Wellington, the capital, 1,000 kilometres to the south, and they were led by an eighty-year-old woman with severe arthritis: Whina Cooper.

Dame Whina Cooper was born Hohephine Te Waka in 1895, the daughter of a local Maori leader. The family had status, but little else. In common with most Maoris, Cooper's family was poor.

She did receive an education – walking six miles every morning to primary school – and, later, attending a college for Maori girls with financial help from family friends.

At sixteen she rejected an arranged marriage, upsetting many in her family and her community, and at eighteen she offered further proof of the rebellious instincts that would, years later, bring her to national attention.

A local white farmer had begun draining mud flats for grazing land that the local Maori claimed as their own. Hohephine led a group of young adults who filled in the drains as fast the farmer could dig them.

Whina Cooper campaigned for Maori justice and land rights from then on, but she also worked, briefly, as a teacher and as a housekeeper at a Catholic presbytery. (She was a devout Catholic and considered becoming a nun.) She was president of the North Hokianga Rugby Union and she became a wife, mother and grandmother.

But it was only after the death of her second husband, Bill Cooper, in 1949, when she moved to Auckland, that she began to become a more serious threat to New Zealand's white political leadership.

Her argument – that the original inhabitants of New Zealand, the Maoris, had been robbed by the pakeha (New Zealanders of European descent), who arrived hundreds of years later – was made all the harder to ignore by the fact it was being made by this passionate, dignified woman.

Cooper founded effective projects including, in 1951, the Maori Women's Welfare League. But the fundamental problem of land ownership was no closer to being solved. And so in 1975, at the age of eighty, she decided to show white New Zealanders how strongly the Maori felt.

Her march from Te Hāpua to Wellington might have started with only fifty people, but by the time they reached the capital, the protestors' numbers had swelled to around 5,000. When Whina Cooper reached Parliament, a month after setting off, everyone was watching.

Not all Maoris supported Whina Cooper's tactics. Some saw Cooper's approach as too gentle. She was criticised for accepting honours (she was made a Dame in 1981) offered to her by the white system. (Cooper would say that honours made white politicians more inclined to listen to her.) But no one could doubt her bravery or seriousness of purpose.

When she died in 1994 at the age of ninety-eight, more than a million New Zealanders followed her funeral on television.

Dame Whina Cooper had become a hugely important national figure. But she might have set herself an impossible task. It is still the case today that only about five per cent of New Zealand's land mass is in the hands of the Maoris.

John Glenn

In February 1962, John Glenn sat alone atop a Mercury-Atlas D rocket in the *Friendship 7* spacecraft. The Atlas rocket was a modified intercontinental ballistic missile. He had every reason to be fearful.

A few years earlier, Glenn and the other Mercury astronauts had been taken to watch a test launch of the Atlas rocket. Like the four previous test launches, it ended badly; the rocket exploding a minute into flight.

If the Atlas rocket did deliver Glenn to orbit, he would then face a new series of potentially mission- and hence life-ending problems. If he survived, however, he would become the first American to orbit the Earth.

The United States was at that point losing the space race. Gagarin and Titov had already achieved orbit. Glenn's mission would, if successful, prove NASA could catch the Soviets.

There were enormous challenges. After orbiting the Earth three times, an indicator showed the *Friendship 7*'s heat shield had failed, meaning it was very likely the spacecraft would be destroyed on re-entering the Earth's atmosphere. There was only one way to find out.

Glenn didn't burn up on re-entry: it was the indicator, not the heat shield, which had failed. Glenn returned safely to Earth and became – instantly – a national and international hero.

The coincidence of geopolitics, the mass uptake of TV, a worldwide infatuation with space and the fact the mission had appeared close to tragedy, made John Glenn not only the most famous and admired man in the world, but perhaps the most famous and admired man the world had ever seen.

John Glenn was born in Ohio in 1921. He was a Marine Corps pilot in the Second World War and Korea and was extravagantly decorated. He had demonstrated, many times, that he was well endowed with what Tom Wolfe called the 'right stuff'.

He did what all the best – or at least the most supremely self-confident – military pilots wanted to do: he became a test pilot. Back then, testing cutting-edge fighter jets was very likely to get you killed. But then crashing and dying, regardless of fault, was a sign that perhaps the pilot in question didn't have the 'right stuff' after all.

Glenn survived and in the late 1950s was selected by NASA, after an exhaustive, almost ludicrous, elimination process, to become one its first astronauts: one of the 'Mercury Seven', or 'Astronaut Group 1'.

Glenn was not like the others. It was generally important to fighter pilots – and all the Mercury astronauts were military pilots – not only to fly faster and higher than their peers, but to do so without seeming to take it too seriously.

If you could successfully complete a mission after twenty-four hours spent drinking, womanising and racing cars in the desert, your claim to the 'right stuff' would be all the stronger.

But Glenn – this 'freckle-faced son of Ohio', as the *New York Times* called him – went to church every Sunday. He jogged each morning before the others had surfaced. He lectured the group on how they should behave and reminded them, regularly, that the world was watching.

So perhaps it's not surprising that it was Glenn who, despite never going to the Moon (it's said Kennedy asked NASA not to send him, because his celebrity was so useful), was seen as the astronauts' leader and the one for whom a political career was most likely.

Glenn became a US Senator in 1974, after a couple of false starts, and kept his Ohio seat for twenty-four years. He was a reliable liberal and he was effective. In 1978, he wrote the Nuclear Non-Proliferation Act – arguably a greater achievement than his space flight.

His life to this point had been one of breathtaking achievement. But there was a further accomplishment to come – one which qualifies him for inclusion in this book.

Shortly before leaving the Senate, in 1998, Glenn lobbied NASA to give him one more spaceflight. It seemed a crazy idea. He was seventy-seven years old! But somehow he persuaded them.

NASA claimed Glenn's flight might present a useful opportunity to observe the effects of weightlessness on an ageing frame. Many suspected NASA was more interested in exploiting a PR opportunity. There was also a suggestion that perhaps NASA owed Glenn. He had done everything that had been asked of him, but he had been denied the greatest prize: a Moon shot.

So on 29 October 1998, Glenn sat, with others, in the space shuttle *Discovery*, strapped to a set of solid rocket boosters, looking forward to a nine-day mission.

John Glenn, the first American to orbit the Earth, became the oldest person ever to go to space. He was also the last surviving Mercury astronaut when he died, in 2016, at the age of ninety-five.

Flora Thompson

Flora Thompson left little evidence of her life beyond her own books. She was a life-long outsider, a watcher of places and people, so perhaps she was not in the habit of expecting others to take much interest in her. There are few letters, no diaries; a modest archive.

But her trilogy, *Lark Rise to Candleford*, is an enduring British classic and still popular many decades after its first publication. It took Flora most of her life to work up to writing it. She was left with very little time to enjoy the rewards that it brought once she had.

She was born Flora Timms in Oxfordshire in 1876. Her father was an aspiring sculptor and her mother had a passion for music and the arts. This might have seemed an unwelcome pretension at a time when most people the family knew were struggling simply to find enough to eat.

Flora was considered plain and bookish. It was clear that she would need to find employment, which she did, becoming a Post Office clerk at the age of fourteen.

She was sent to work in Grayshott (at the time a literary hotspot, counting George Bernard Shaw and Arthur Conan Doyle among its residents) and later to Bournemouth where, in 1903, she met and married John Thompson, a Post Office colleague.

All the while she wrote. She won an essay competition and began to publish short stories in the *Ladies Companion*. Her stories were based on her own life and frustrations, and on her detailed observations of others.

She discovered that stories depicting the rural life of her childhood seemed to provoke the most satisfying response from readers and so in 1935, by now almost sixty, she started writing *Lark Rise*. She was accused by some of sentimentalising the impoverished rural England she had grown up in, but her writing was far sharper than that.

The nature writer Richard Mabey argues in his biography of Flora Thompson, *Dreams of the Good Life*, that she was no amateur or 'hedge-scribe'. She was, he writes, 'a sophisticated and imaginative writer, involved in a more complicated business than straightforward autobiography.' Melanie McDonagh suggests in the *Spectator* that Thompson wrote 'the kind of social history only a woman could write'.

Lark Rise was the first of three books, which were published together as a hugely popular trilogy in 1945. Flora was by then sixty-nine years old. She died only two years later.

There is a sense that Flora Thompson was underestimated throughout her life, and perhaps even in death, too. She was certainly slow to get started and success came right at the very end of her life. A plain Post Office clerk in pre-war Britain, she successfully re-invented herself as a professional writer. But, even so, there is a sense she remained on the edge of her own story.

She surely wouldn't have allowed herself to imagine that her books would be read by millions – or that they would still be read seven or so decades after her death.

Colonel Sanders

Today Colonel Sanders is a global logo. His familiar goateed face smiles over 23,000 restaurants in 136 countries, which together made sales of $26 billion in 2018 alone.

This posthumous ubiquity is a dubious distinction (any number of vegans, vascular surgeons – not to mention chickens – might question it). But there is at least one good reason why we *should* admire the Colonel.

Harland Sanders was born in 1890 to a farming family in Indiana. After his father died in 1896, and his mother found work peeling tomatoes in a canning factory, the future fried-chicken tycoon was left in charge of his siblings. He spent long hours in the kitchen, where he showed early promise.

When an uncharitable stepfather moved in, the twelve-year-old Harland moved out. He soon left school and found himself alone in the world. But he wasn't entirely unprepared. For one thing: he was already a great cook.

Also, his mother had imposed – among other things – the idea that Harland should stay away from booze and cigarettes and knuckle down instead to hard work. (Her views on saturated fat are not recorded.)

Over the next four decades or so, Sanders accumulated a strikingly varied CV: farmworker, fireman, insurance salesman, tyre-seller, steamboat ferryman and, finally, gas station manager. It was the last of these roles that put him on the path to finger-lickin' greatness.

Sanders – by now an honorary Colonel of the State of Kentucky (a distinction that is, in truth, fairly easily achieved) – began offering fried chicken to hungry motorists passing through his gas station. Soon his reputation stretched up and down the highway and so he built a restaurant.

In the kitchen of that restaurant, Colonel Sanders made a gastronomic discovery that would echo around the world (eventually). He found that by frying chicken inside a pressure cooker, he could get faster and juicier results.

But it wasn't enough. When a new interstate highway was built, leaving Sanders' restaurant stranded on the old and soon-forgotten highway, the business went under.

Here's where we might allow ourselves to begin admiring the ageing Colonel. Because now he was in his mid-sixties. And he didn't have a lot to show for it: a car; a white suit; some modest savings; a now-legendary $105 social-security cheque; and he had to admit, a killer way with chicken.

Could he put these ingredients together, shake them up like a bag of seasoning, and produce a zinger of a business model that would make millionaires of him and, in time, thousands of others? You know the answer.

He travelled state to state, town to town, restaurant to restaurant, looking for proprietors willing to franchise his method. He often slept in his car. He kept a pressure cooker in the boot and pulled it out everywhere he went, ready to demonstrate how good his fried chicken could taste.

Such persistence is admirable by any standards, but it is extraordinary for a man who might have considered himself, back then, to be not only at the end of his working life, but the end of his life.

By 1963, at the age of seventy-three, Sanders had six hundred Kentucky Fried Chicken franchisees in the US and Canada. In 1964, somewhat reluctantly, he sold the enterprise for $2 million, some shares, and a promise that his recipe would never be altered.

According to a *New Yorker* profile published in 1970, Sanders then spent his time touring the US – by now an instantly recognisable celebrity – promoting KFC.

Sanders lived the rest of his life in Kentucky. He lived modestly by the standards of a rich man. But then, he wasn't *that* rich – future custodians of KFC would become far richer – and he was old; he didn't see much need for fancy things.

When he wasn't touring the United States encouraging his fellow Americans to eat fried chicken, he was worrying – perhaps not unreasonably – about standards. As the *New Yorker* put it: 'During most of his waking hours, the Colonel is fretting that somewhere a careless franchisee is overcooking or undercooking his chicken.'

They might not have been then. Someone, somewhere almost certainly is now. But it no longer matters to the Colonel. He died in 1980 at the age of ninety.

Very few people watching an old man struggling to pull a pressure cooker out of a battered car – his suit crumpled from having slept in it – can have guessed that they were looking at the birth of a global brand.

Baroness Trumpington

Baroness Trumpington was born Jean Campbell-Harris in 1922. That sentence hides the secret of the late success for which she is included in this book.

Her mother was an American heiress, thanks to the success of Indestructible Paint Ltd, her grandfather's company. Rich and well-connected, the young Campbell-Harris moved – albeit obliviously, according to her autobiography – in highly fashionable circles. Her parents counted the Prince of Wales among their friends.

But in 1929, the Wall Street Crash revealed the old man's paint company to be anything but indestructible and the family lost a good part of its fortune. Somehow, quite mysteriously, life seemed to carry on much as it had before: Jean was sent to a succession of elite schools and the Prince of Wales was still in the address book.

Then war came. Jean was sent to Churt, Surrey, to work as a 'land girl' on David Lloyd George's farm. The former prime minister was, of course, a family friend – though he seems not to have let that inhibit his behaviour towards the seventeen-year-old girl.

He would ask her up to the house and order her to stand against a wall for measuring. He would pull a tape from his pocket and set about recording the future Baroness Trumpington's generous dimensions. She seems to have taken this (and many other strange and unwanted advances from ageing establishment figures) in her stride.

Then came Bletchley Park. Campbell-Harris spoke German and French, which made her useful. She was also posh, making her trustworthy – in the eyes of the kind of people who ran such places in 1940s Britain.

Bletchley Park was a code-breaking operation that played a huge part in the Allied victory. Her job was to translate intercepted German naval communications. Throughout this period, she continued to see off the approaches of men young and old; or, as she put it, 'we'd hang on to our knickers as hard as we could.'

After the war, Jean Campbell-Harris met Alan Barker, who would become an eminent headteacher, and soon became Jean Barker. She settled into life as a headteacher's spouse and new mother but, after a time, began to feel she needed something else to fully 'occupy my energies'.

She went into local politics – Conservative, naturally – and served as a local councillor and as mayor of Cambridge. Her energies were still unsated, however, and in time she set a course for London, seeking selection as a Tory candidate for Parliament.

She was rejected by local Conservative associations – twice – and by the late 1970s, it seemed to both Barkers that their careers had stalled. Alan reconciled himself to the disappointing truth that he would never be headmaster of Eton. Jean reluctantly conceded that she probably wasn't destined for the Commons. But then Margaret Thatcher intervened.

In 1980 – egged on, perhaps, by mutual friends – Mrs T elevated Mrs B to the House of Lords. The newly minted Baroness took the name 'Trumpington' from the Cambridgeshire ward she had once represented.

It was an inspired choice – a magnificent example of successful rebranding. Baroness Trumpington had adopted an unforgettable persona perfectly calibrated to suit the public figure emerging from behind the name.

Large, fearsome, seemingly as British as a slab of spotted dick – and now armed with a faintly ludicrous but imposing name – Baroness Trumpington began to be noticed. Her fellow Lords, certainly, could hardly miss her.

As she sailed into her sixties, she found herself in government. She was a junior minister in the health department, despite being a militant smoker, and later served in the agriculture department. She became the oldest woman ever to serve in government. In 1992, John Major asked her to step down from her post. The Baroness burst into tears and Major relented.

In the Lords, she was noted for her forthright style. In 2011, she was caught on camera sticking two fingers up at Lord King, who had made what she regarded as a disparaging remark about her age.

The following year, aged ninety, she appeared on the satirical TV show *Have I Got News for You* – their oldest panellist ever. Trumpington, of course, more than held her own. In 2014, she published a successful autobiography, *Coming Up Trumps*, and she appeared in a TV programme about fashion for the elderly.

By now Trumpington was, if not exactly at the centre of popular culture, not as far away from it as one might expect. She had turned her style, her name and her age into a compelling 'media offer' – and she had a great deal of fun with it in her final years.

It would be overstating it to say Baroness Trumpington changed the world. But her story does prove that even in your sixties, when your friends and contemporaries are contemplating retirement, your career might be about to take off.

Baroness Trumpington – the former Jean Campbell-Harris – died in 2018 at the age of ninety-six.

Taikichiro Mori

When he died in 1993 at the age of eighty-eight, Taikichiro Mori was thought to be the world's richest man. He had made his fortune in Tokyo property and his worth was roughly double that of Microsoft's Bill Gates – then the richest American.

Silicon Valley was about to change everything, of course, but it wouldn't have mattered a great deal to Mori. He seemed bemused by his super-rich status and the attention it brought.

His lifestyle didn't reflect his wealth and he would no doubt have viewed the uncontrolled consumption of today's digital rich and plundering oligarchs as obscene. He was, one might say, a very Japanese billionaire: modest, thoughtful and exceptionally clever.

Mori was born in 1904 to a rice trader and property owner. The family business was centred on the Toranomon neighbourhood in the Minato ward. Mori would recall the area with fondness.

'In my childhood, the roads were very narrow,' he said, 'and just like a snake who swallowed an egg, they would widen out in places, and children would play hide and seek and spin tops without danger.'

The neighbourhood would be dramatically reshaped three times during Mori's life – the last of these, his own work.

In 1923, a huge earthquake destroyed great swathes of Tokyo, killing at least 100,000 people. In 1944 and 1945, US aircraft firebombed Tokyo repeatedly, destroying much of the city once more, and again, killing at least 100,000 people and displacing a million more.

Mori's first career was in academia. He worked steadily and, by the age of fifty, had become head of the School of Commerce at Yokohama City University. But then, a few years later, he complained of boredom, and decided to try his hand in another role before he was too old. In 1959, he formed the Mori Building Company.

Mori turned first to the Toranomon neighbourhood. He worked carefully and slowly, apparently treating sceptical residents in an attentive manner and eventually gaining wide support for his modernisation plans.

Over the next two decades, Mori turned the area into a modern cityscape. Recalling the fragility of Tokyo's traditional wooden architecture, Mori built with steel and concrete.

He felt that modernising Tokyo was an essential element in renovating the economy of all Japan.

By the time he died in 1993, Mori's company controlled over eighty important buildings in central Tokyo. Property prices were firmly in 'bubble territory' around that time – flattering his paper worth – but even allowing for a price correction (one he anticipated and even thought desirable), Taikichiro Mori had become almost unimaginably wealthy. His achievement was all the more extraordinary, given he turned to business at the age of fifty-five.

In 1989, Reuters reported that Mori still lived in much the same manner as he had when he was a moderately successful academic. Employees who saw his apartment were astonished by its ordinariness. He wore a simple black kimono (he found ties uncomfortable) and appeared to have very little interest in the typical trappings of success.

'Our household runs just like any other Japanese household,' he said. 'I live a life that leaves me content and happy. I was used to living like a professor and I wasn't about to change.'

Peter Roget

The story behind Peter Roget's thesaurus, his most enduring contribution to the world, is surprising, remarkable and unexpected.

He began listing synonyms as a kind of therapy. For Roget, such obsessive attention to order and classification provided an escape from a life that was frequently blighted by tragedy. What's really surprising, though, is that the thesaurus might not have been Mr Roget's greatest work. He was an exceptionally inventive, ingenious and resourceful man.

Born in 1779 in London to a Swiss vicar and a jeweller's daughter, it quickly became clear that Peter Roget was unusually bright. But when his father died in 1883, the family struggled.

They moved frequently and relied on the kindness of the wider family, until they found their way to Edinburgh where, at the age of only fourteen, Peter Roget began studying medicine. He was still a teenager when he graduated from Edinburgh University and became a doctor.

Peter Roget's mind roamed well beyond medicine. He made important contributions to biology, optics and mathematics. His observations of optical illusions in 1820 contributed to the development of the zoetrope and hence the motion picture.

He invented the 'log–log' slide rule for calculating the roots and powers of numbers (a tool, device or apparatus that remained in use until the invention of the calculator) and he published papers on the effects of laughing gas, among other things. He was secretary of the Royal Society for twenty-one years.

But throughout all this, he struggled with his own depression and that of his controlling mother and unhinged sister. His uncle, Sir Samuel Romilly, a distinguished member of parliament who had campaigned against the slave trade, took his own life by cutting his throat. Peter Roget was in the room when it happened.

Making lists soothed Roget when nothing else could. It might have been a kind of compulsion – perhaps even a condition in its own right – but it was an immensely useful one.

Roget had begun collecting words early in his career and he continued to add to his great list throughout his life, disappearing into the project for days or even weeks at a time, when depression, despair or melancholy threatened to overwhelm him.

When he retired from medicine, Roget devoted himself full-time to his thesaurus. He worked on it for a further twelve years before it was finally published in 1852. Peter Roget was by then seventy-three years old. His *Thesaurus of English Words and Phrases* has been in print ever since.

Roget's wasn't the first attempt at such a compendium, compilation or anthology, but it was the best. His book is still considered an indispensable reference today.

Some criticised Roget for making it possible to choose words without thinking. But almost every English-language writer in the last 150 or more years has – surely – referred to Roget's magnum opus.

Peter Roget kept on adding to the book, retreating into the comforting order of his thesaurus whenever he could, until his death, demise or extinction in 1869 at the age of ninety.

Annie Jump Cannon

Annie Jump Cannon joined the faculty of Harvard University when she was made a professor of astronomy in 1938. She was seventy-five years old and had been working in astronomy – and making an extraordinary contribution to the field – her entire life.

Had she been a man, Annie Jump Cannon would almost certainly have joined Harvard's professorial ranks several decades earlier; that she made it all, however late, was a rare achievement.

Born in 1863 in Delaware, Annie Jump Cannon was introduced to stargazing by her mother, Mary, who had studied astronomy at a Quaker school. The pair spent many evenings together learning the constellations.

That, and a clear talent for mathematics, persuaded Annie's parents to encourage her further education. She went to Wellesley College and thrived there, graduating with a degree in physics.

Annie was also an early adopter of photography, contributing images taken in Spain to exhibitions. As a young woman, she contracted scarlet fever and lost much of her hearing.

She sometimes said that near-deafness, together with an almost complete lack of other 'distractions' in her life (she never married and had no children), were the secrets to her success; she was dedicated completely to her work, and she had an astonishing appetite for it.

Annie Jump Cannon was able to bring her interests in photography and astronomy together in studying the spectroscopic analysis of light; understanding the composition of stars by observing radiation passing through a prism.

At the time, Edward C. Pickering, director of the Harvard College Observatory, was attempting to catalogue every visible star in this way. It was an epic task, which required great diligence and patience. He employed a number of women as assistants, including Annie Jump Cannon.

Annie not only proved herself faster and more accurate than anyone else engaged in the project, she devised a new and better method of classifying 'stellar spectra' – one that is still, with some adjustments, in use today.

The Harvard catalogue was released as the *Henry Draper Catalogue* over several editions from 1918 onwards, eventually charting nearly 300,000 stars – each of them individually analysed, many of them by Annie Jump Cannon.

Finally, in 1938, in advanced old age, she was appointed a professor. Over the course of Annie Jump Cannon's life, women slowly came to be, if not accepted, at least tolerated in the sciences – a realm, until then, preserved almost exclusively for men. When she died in Cambridge, Massachusetts, in 1941, there was still a long way to go. (Maybe there still is.)

But while she had to work harder than any male peer, make far more sacrifices and many fewer mistakes, Annie Jump Cannon did find a way to make science her life. In doing so, she undoubtedly inspired many others to follow her lead.

When she was born in 1863, the American civil war was raging and Abraham Lincoln was president. When she died, Franklin D. Roosevelt was president and Japan was months away from its attack on Pearl Harbor.

She saw profound change in her lifetime. The US had moved firmly into the modern era. And it was led there by scientists – like her.

Clint Eastwood

Should Clint be in this book? Maybe. He was late to success by the standards of leading men in 1950s Hollywood. But he wasn't that late.

Before he found his way to LA (which wasn't far – he was born in San Francisco), Clint Eastwood was drafted into the US Army during the Korean War (he saw no action), survived a plane crash (swimming two miles to safety), dug out swimming pools, and worked on a golf course.

He then spent several years playing instantly forgettable bit parts in long-forgotten films. Finally, in 1959, at the age of twenty-nine, he found fame as the cowboy Rowdy Yates in the hugely successful TV serial *Rawhide*.

What is remarkable about Clint Eastwood, making him unique in Hollywood – and worthy of inclusion here – is the longevity of his career both as an actor and, from the 1970s onwards, as a director.

He was born in 1930, but neither age nor success (he has amassed more wealth than he could ever hope to spend) have slowed him down. The British Film Institute credits Eastwood with creating 'one of the best bodies of work in American cinema'.

In the early 1960s, *Rawhide* made Clint Eastwood a household name, but the films that followed – the iconic 'spaghetti westerns' *A Fistful of Dollars*, *For a Few Dollars More*, and *The Good, the Bad and the Ugly* – made him an international star.

His dubious heroics – he was the good guy, but only just – suited the mood of the time. Clint Eastwood became one of the fortunate few whose early work is so good they never outlive it.

In 1971, at the age of forty-one, Clint Eastwood made his directorial debut with *Play Misty for Me* and he's never stopped. He has made movies (starring in many of them) more or less back-to-back ever since.

Film people will tell you directing is an exhausting, physical process, and that few can keep going beyond their sixties. Clint has no time for that. 'Everybody thinks making films back-to-back is a big deal,' he has said, 'but they did it all the time in the old days.'

His films include some great ones – *The Outlaw Josey Wales*, *Unforgiven*, *Million Dollar Baby* – and some duds. But Clint doesn't appear to spend a lot of time dwelling on it either way. He is, it's said, exceptionally efficient: his films are made on time and never go over budget. But perhaps that's no surprise.

Can you imagine being the prima-donna actor refusing to come out of your trailer on a Clint set? Or the lighting guy who says he needs more to time to set up the take? Clint would have only to curl his lip almost imperceptibly and the shoot would surely be back on track.

As Joe Queenan put it in the *Guardian*: 'Unlike sensitive auteurs, who will take a few years off to contemplate their next project, Eastwood has not stopped making films since his debut in 1971. This has been one long career. Eastwood has outlasted all of his notable contemporaries.'

Clint Eastwood has combined his long film career with a sideline in red-in-tooth-and-claw conservative politics. He says things right-wing Americans love to hear, such as 'I have a very strict gun control policy: if there's a gun around, I want to be in control of it!' He was mayor of his hometown Carmel in the 1980s and he appears to revel in being a rare conservative in an industry dominated by liberals.

Between his unending work, and his hobby antagonising left-wingers, Clint Eastwood seems to be enjoying himself as much now as he ever did. 'Ageing can be fun if you lay back and enjoy it,' he once said.

And perhaps that's why he should be in this book. He didn't let his early success diminish his appetite for more. It never seems to have occurred to him to slow down. Just like the cowboys in his movies, it seems Clint Eastwood won't give up until he's carried away in a box.

Melchora Aquino

In August 1896, several hundred members of the Katipunan, a secret Philippine patriotic organisation, descended on Balintawak, near Quezon City.

Their existence had been discovered by the Spanish colonial authorities earlier that summer and now the Katipunan, led by Andrés Bonifacio, faced a difficult decision. Spanish troops were searching for them. Should they turn themselves in, or fight?

They holed up at the farm of a local woman called Melchora Aquino. Aquino had grown up in poverty. Her husband had died young and she had brought up her large family alone. But now she had a granary, provisions and animals, and decided that she would offer all of this to these young men in hiding.

She fed perhaps a thousand people over those few days, giving medical attention as best she could, and saying prayers. Her effort was all the more extraordinary because she was eighty-four years old, which, at the end of the nineteenth century in the Philippine back country, was almost unimaginably old.

It was there, on her farm, that Andrés Bonifacio resolved to start the Philippine Revolution. The Katipunan tore up their Spanish-issued identity papers – a sign of commitment to the struggle – let rip their battle cries and resolved to attack Manila within days. Melchora Aquino has been known ever since as the 'mother of the revolution', or the 'grand woman of the revolution'.

Spain had controlled the Philippines for over 300 years by this time. The revolution would last for three years. Many were killed on both sides. When Spain lost the 1898 Spanish–American war, it was forced to cede control of the Philippines to the United States.

Philippine revolutionaries soon turned their guns on their new colonial masters, but the Philippines didn't achieve full independence till 1946 – fifty years after Melchora Aquino helped prepare the country's nascent revolutionary army for war.

When Spanish soldiers discovered Aquino's support for the Katipunan in 1896, she was interrogated. But she refused to divulge any information and was sent to a penal colony on Guam.

She was forced to live there for seven years. When she was brought back to Manila in 1903 aboard a US warship, she was ninety-one years old. She apparently lived for a further sixteen years.

Melchora Aquino is an important figure in Philippine history. She has appeared on banknotes and her remains were moved in 2012 to a purpose-built shrine in Quezon City. Aquino didn't go looking for history, it came to her. But when it did, the 'mother of the revolution' – despite being in advanced old age – was more than ready for it.

Frank McCourt

Frank McCourt was born in Brooklyn in 1930, but grew up in Limerick after his Irish parents retreated 'home' to escape their tough life in New York.

Life in Limerick would prove even harder. McCourt's father, Malachy, moved to England in search of work and never came back. His mother, Angela, raised her four surviving children (three died in childhood) in a rat-infested slum, making ends meet through – among other things – sleeping with her cousin in return for rent.

At nineteen, McCourt made the journey back to New York, hoping he could make a better go of it there than his parents had. He had little education, no money and few contacts. But though it would take time, McCourt did find salvation in America.

The Korean War gave him his break. He was drafted into the US Army, but instead of being sent into combat, he was posted to Germany to work a desk job. That meant formal Army training in useful office skills such as typing and three years later, after completing his service, the chance to study for a degree. McCourt chose English and education and set himself up for a career in teaching.

He taught English literature and creative writing to hard kids in tough New York public schools; kids who had little interest in reading the classics and, perhaps correctly, judged they would never need to.

McCourt's teaching career was important and satisfying and he was probably very good at it. But, like plenty of English teachers, he felt he had a novel in him. As it turned out, he had something arguably much better.

In the mid-nineties, now well into his sixties and having retired from teaching, he wrote a memoir, *Angela's Ashes*, which describes his bleak childhood with stunning lyricism.

Angela's Ashes was published in 1996 and was an instant sensation. It was translated into seventeen languages, made into a Hollywood film, and went to the top of the bestseller lists in over twenty countries. It won him the National Book Critics Circle award and the Pulitzer Prize.

Why did it take McCourt so long to write his story? He had struggled for decades with the question of what to write. He had written short stories, some of which were published, but his breakthrough came when he stopped trying to imagine others' experiences and presented his own, in his own voice.

Further memoirs followed – '*Tis* in 1999 and *Teacherman* in 2005 – both of which also met with critical and commercial success.

There was a backlash in Ireland, where many felt McCourt's dark depiction of Limerick was exaggerated and unfair, but no one could dispute the quality of McCourt's writing. It seemed extraordinary that such a talent could have remained more or less hidden for over six decades.

Writing had finally made McCourt rich, famous and widely respected, particularly in the United States. He is said to have worn his success well, not letting it go to his head, as it might have done had he achieved all this as a younger man.

But having turned his whole life experience into books, there was nothing left for him to write. There was no novel after all.

Frank McCourt died in 2009 at the age of seventy-eight.

Lise Meitner

Lise Meitner was sixty-one years old in 1939 when, together with her nephew Otto Frisch, she calculated a theoretical explanation for nuclear fission – the splitting of the atom.

By 1942 the United States had established the Manhattan Project, an all-out effort to build a useable nuclear weapon. In 1945, the US dropped 'Little Boy' on Hiroshima in Japan and, three days later, 'Fat Man' on Nagasaki. Both towns were destroyed and hundreds of thousands were killed.

'You must not blame us scientists for the use which war technicians have put our discoveries,' Meitner once said. She, like Einstein, had refused to work on the Manhattan Project. She abhorred the violence her work helped make possible.

Meitner was born in Vienna in 1878, one of eight children. She showed an early aptitude for mathematics and her father saw no reason why his daughter should not aim at least as high as his sons. She was privately tutored and eventually graduated from the University of Vienna with a PhD in physics.

Max Planck invited her to Berlin to further her research. Despite the support of such an eminent academic, however, and her own obvious ability, she had to contend with endemic sexism.

Shortly after reaching Berlin in 1907, she met Otto Hahn, who would be her research partner for the next thirty years. She was treated for a time as an unpaid assistant. But most of the faculty – and certainly Hahn himself – knew full well she was at least his intellectual equal.

Their research partnership flourished and the pair made rapid progress uncovering the physical properties of radioactive elements. Meitner's reputation continued to grow among physicists well beyond Germany and Austria.

When the First World War came, Meitner was deployed to the Austrian front to work as an X-ray technician. There she saw the consequences of war and knew she wanted no part in it.

But within decades another war loomed. And as a Jew in Berlin – albeit a non-practising one with skills the Nazis might covet – Meitner was in danger. Her colleagues in the wider European physics community knew it, too.

In July 1938, she was helped to escape Nazi Germany. It was an extraordinary operation, masterminded, among others, by the great Danish physicist, Niels Bohr. Meitner was accompanied for much of the journey by a Dutch physicist, who sat a few seats away from her on the train from Germany to the Netherlands. She had no idea he was there. Her final destination was Sweden.

It was in Sweden, walking in the woods with her physicist nephew, Otto Frisch, that Meitner figured out how to understand the 'fission' that Otto Hahn had observed in experiments in Berlin.

Hahn had isolated fission, the splitting of the atom, but Meitner had supplied the explanation. Their long association had yielded something of almost incalculable importance.

But the partnership was not seen as such by everyone. In 1944, Hahn was awarded the Nobel Prize in Chemistry for his 'discovery of the fission of heavy nuclei'. Neither the Nobel Institute nor Hahn mentioned Lise Meitner's essential contribution.

Meitner eventually left Sweden to live in Cambridge with Otto Frisch. She died there in 1968. Frisch had her gravestone inscribed with this epitaph: 'A physicist who never lost her humanity.'

Henry Ford

Henry Ford was born in Dearborn, Michigan, in 1863, to a farming family. His future was perfectly clear, or so his family thought: he would take over the farm. But Ford, of course, had other ideas.

His story is familiar – he is the archetypal self-made man – but it remains extraordinary both for the number of calamitous setbacks he overcame, and the truly astonishing scale of his success when he finally made it.

Ford's education was apparently rudimentary, but it's clear he was an intuitively talented engineer, perhaps a genius, right from the start. As a child, he would take apart and repair neighbours' watches and clocks. He would build water wheels and steam engines, learning as he went by trial and error.

In 1879, he went to work as an apprentice building railroad cars. He held several other junior engineering jobs, worked in factories, cut and sold timber, and at one point he serviced

Westinghouse steam engines like the one used on his own family farm, where he also worked, briefly, despite his desire to escape it.

Ford bounced around positions like these for over ten years, before he found a job working nights at the Edison Illuminating Company in Detroit. He had no experience of electricity, but he would learn – as he had learned from all his previous roles.

He did well at Edison and was promoted to chief engineer in 1896, when he was thirty-three years old. By now, in his own time, he was indulging what had become his chief interest: devising and building 'horseless carriages', or motor cars. He invented a contraption called the Quadricycle, and a number of other innovative but undesirable powered vehicles.

Along the way, however, Ford had acquired some useful supporters, including Edison himself, and in time he was able to marshal this network and its capital behind his ideas and his ferocious ambition.

In 1899, he founded the Detroit Automobile Company. Ford was now thirty-six. A white male born in the United States in 1863 would have had a life expectancy at birth of fewer than fifty years. So although Ford was hardly ancient, and certainly not old by modern standards, he would not have been regarded as a young or even middle-aged man either.

He had left it late to change the world. And he failed. The automobiles produced by Ford's first company were unreliable and too expensive. The firm collapsed two years later.

Almost immediately, though, Ford found new investors to try again and in 1901 he established the Henry Ford Company. But Ford fell out with his backers and left the company, which would be renamed Cadillac. Finally, in 1903, he created the Ford Motor Company. And this time, he got everything right.

The Ford Motor Company launched the Model T in 1908, when Ford was forty-five years old. Henry Ford had not only designed the car to be cheap and reliable, and as easy as possible to repair, but in time he designed the entire factory and supply chain behind it to be breathtakingly efficient.

Ford did not invent the concept of the assembly line, but he took the idea and applied it to his business on a scale and with a ruthless attention to detail never seen before. The Model T was a huge success. Pretty soon, at least half the cars in America were Fords.

Ford would build the biggest factory in the world, where every element of a car's production was brought together – glass-making, metal work, tyre production – so that he had control over every part of the process. Ford cars were well-made and affordable and within the reach of ordinary Americans.

Ford retained control of the company (he had bought out all his minority investors in 1919 for $105 million) until his death in 1947 at the age of eighty-three.

He is understandably revered as a great figure in American industrial history. But he was certainly not perfect. He paid his workers well – in part, it's said, so they could buy the very cars they made – but he had a notably uncompromising approach to industrial relations. He also pursued a strange sideline in rabid antisemitism, publishing hateful conspiracist nonsense in books and in his own newspaper.

Ford might also have been lucky. He was in the right place at the right time. Had he been born twenty years earlier, or twenty years later, or in Missouri rather than Michigan, we might never have heard of him.

He was there on the eve of the era of the automobile, with the right skills and the right ideas and in the place where it would happen. But then, of course, the same could be said for countless others. And it was Ford who, through tireless persistence, came to stand above them all.

Hannah Hauxwell

Hannah Hauxwell was destined to live a life of extreme obscurity. She was born in 1926 to a North Yorkshire farming family. The farm – Low Birk Hatt – was tenuous. It occupied a lonely, windswept 78 acres alongside the Pennine Way. There were no modern machines or conveniences.

By 1961, Hannah's parents and Uncle Tommy, to whom management of the farm had been passed after her father's early death, had all passed away. Hannah was entirely alone. She could go weeks without seeing another living soul. She tended a single milking cow and scratched out a life on £5 a week.

But she was, she said, perfectly happy. She appears only rarely to have questioned her isolation. She endured ferocious winters with little complaint – not that there was anyone to complain to – and she felt entirely at home alone in her deteriorating farmhouse.

In 1970, the *Yorkshire Post* featured Hannah in an article entitled, 'How to be happy on £170 a year'. Not long after that, she was 'discovered' by Barry Cockcroft, an enterprising TV documentary maker from Lancashire.

Hauxwell was forty-seven in 1973 when Cockroft's film about her, *Too Long a Winter*, was broadcast in the UK on ITV. You might think forty-seven is hardly old enough to qualify for inclusion in a book about people who came to prominence late in life. But no doubt as a consequence of a life spent farming alone in the freezing Dales, by then Hannah Hauxwell was an old woman. She had white hair and looked at least twenty years older than she really was.

The film was an instant classic and made a huge impression on viewers in Britain and, in time, far beyond. As the *Yorkshire Post* wrote: 'It is no exaggeration to say that the interview with her that followed broke the nation's heart.'

People from all over Britain, touched by Hannah's poverty and dignity, sent her food parcels, blankets and even money. The sheer quantity of gifts created a serious logistical challenge for ITV.

Her farmhouse was connected to the electricity grid and to running water and she purchased more cows. Hannah was now suddenly – and more than a little improbably – very famous. (This was a time, of course, when Britain had only three TV channels, each of which could command enormous audiences and create stars overnight.)

Cockcroft made further films with Hauxwell. He took her to an awards ceremony at the Savoy Hotel in London for *Hannah Goes to Town* and even to New York City. She met the Pope and she was presented to the Queen Mother at Buckingham Palace.

The 'Old Lady of the Yorkshire Dales', as she had been called since the age of forty-seven, was now in her sixties. Her late fame made her life at Low Birk Hatt more comfortable, but barely more sustainable.

She stayed on the farm as long as she could, despite the near impossibility of making it profitable, and the growing challenge of remaining in such an isolated location in her old age. But in 1988, she finally left, moving to a cottage in nearby Cotherstone.

The move wasn't easy. She had always been a hoarder, through necessity, never throwing away anything that might one day come in handy. But more importantly, she was connected to her farm in the deepest possible way.

The *Yorkshire Post*, the newspaper that had first shone a light on Hauxwell's unusual life, attributed her appeal to the fact she 'spoke with the simple eloquence of one transplanted from another age'.

Her uncomplaining stoicism – and implied rejection of consumerism – made her, in the words of that newspaper, a 'national heroine'. She died in 2018 at the age of ninety-one.

Oscar Swahn

The oldest-ever Olympian is a Brit called John Copley who, in 1948, and at the great age of seventy-three, won a silver medal for his etching of a group of polo players. He was beaten to gold by the Frenchman Albert Decaris and his art deco engraving of a swimming pool. The 1948 London games were the last to include medal categories for the arts and so Copley's record might never be beaten.

If we leave the arts out of it (as the Olympic Committee so wisely chose to do), the oldest-ever sporting Olympian was an extravagantly bearded Swedish gentleman called Oscar Swahn.

Swahn was sixty years old when he entered the 1908 Olympic Games in London, where he took bronze in an event called the 'Men's Double-Shot 100 Metre Running Deer'.

No real deer were harmed in the making of this event. One simply had to shoot at a moving deer-like target, as it was pulled back and forth across a narrow opening. The discipline is no longer part of the Olympics, although it persisted in 'Nordic Shooting Region' competitions till 2004.

The 1908 Double-Shot Running Deer contest was won by Walter Winans, an American with an extensive handlebar moustache and a strong passion for firearms. Swahn would not have to wait long to put the fifty-six-year-old upstart Winans in his place, however. In the single-shot running deer category, Winans finished a distant sixth, while Swahn took gold.

Then came the team single-shot event. Swahn's team-mate was his own son, Alfred, and together these two great Swedish marksmen took gold, beating the British into second place. (British pride in this silver medal was undermined somewhat by the fact that only two teams had contested the event.)

In 1912, Swahn competed on home turf in the Stockholm games. Oscar again took bronze in the double-shot running deer category, before seeing his son Alfred take gold in the single-shot event.

Father and son then successfully defended their 1908 team gold, thrashing their closest rivals, the United States, by twenty points. These were great days for Swedish shooting fans.

Oscar Swahn entered the record books as the oldest-ever Olympic gold medal winner (he was sixty-four years and 258 days old), where he has remained ever since.

In Antwerp in 1920 (the 1916 games were cancelled because

of the war), Oscar and Alfred Swahn again lined up to compete in the team shooting event.

Gold went, no doubt infuriatingly, to Norway, but the Swahns took silver. Oscar was by then seventy-two years and 280 days old, setting another record that has never been (and most likely never will be) broken.

Swahn's 'oldest Olympic competitor record', however, could soon be challenged. The veteran Japanese equestrian rider Hiroshi Hoketsu was seventy-five in 2016, when he planned to compete in the Rio games.

Sadly, Hoketsu had to bow out when his horse became ill. But it's thought he might yet make it to the 2020 games at home in Tokyo. He is unlikely to take a medal, but he would nonetheless become the oldest-ever Olympian at seventy-nine.

Kenneth Grahame

Kenneth Grahame published *The Wind in the Willows* in 1908 when he was almost fifty. At the time, he was secretary of the Bank of England. Over 110 years later, the book remains a popular children's classic and is the reason Grahame is widely remembered.

On the face of it, Grahame is a good example of a talented man who found enduring success relatively late in life. But his story is more complicated than that – and far more tragic.

Grahame was born in 1859 in Scotland. He spent his early childhood on the banks of Loch Fyne in Inveraray. Family life was shattered in 1864, when his mother died from scarlet fever. Grahame's father, a successful lawyer, quickly fell apart and descended into alcoholism.

When it became clear their father could not look after them, Kenneth and his siblings were sent to live with their maternal grandmother in Cookham Dean, Berkshire. She is reported to have been a cold woman. But the boy Kenneth found some comfort in the peaceful willow-lined banks of the River Thames there.

Scroll forward a few years and Grahame appears to have been doing rather well. He flourished at public school, excelling at rugby, and with a little help from what remained of his family, he found a job as a clerk in the Bank of England.

He began writing in his spare time (he had a lot of it; apparently in those days a job at the Bank required very little effort) and achieved significant success with two collections of childhood reminiscences: *The Golden Age* and *Dream Days*.

Under the surface, however, this broad-shouldered and by now luxuriantly moustachioed man, a published author with a serious job, was in some ways remarkably childlike. It's said he was still a virgin at thirty-eight when he met the woman who would become his wife: Elspeth Thomson.

Most biographies suggest the marriage may not have been a particularly happy one. For one thing, Grahame apparently had little interest in sex. But despite that obstacle, the couple produced a son, Alistair, in 1900.

The boy had problems from the start. He was born nearly blind and his behaviour would be described today as very 'challenging'.

Grahame regarded Alistair – he called him Mouse – as a genius, and indulged him. He would write stories for him about the creatures of the riverbank, stories that would eventually become *The Wind in the Willows*.

Meanwhile strange things were afoot at the Bank.

In 1903, a man called George Robinson walked in carrying a rolled-up document and asking to see Sir Augustus Prevost. Prevost had retired as governor several years earlier, so staff helpfully presented Kenneth Grahame to the visitor instead. Unfortunately, Robinson was a 'socialist lunatic', his documents concealed a gun and Grahame was promptly the victim of an assassination attempt.

Luckily, the lunatic missed his mark three times, and Grahame was unhurt. But he was understandably rattled by the episode and pretty soon he left the Bank and retreated to his writing.

The Wind in the Willows was not an immediate hit. Most reviewers panned it. Rather brilliantly, though, Grahame sent a copy to Theodore Roosevelt. The US President had once indicated that he liked Grahame's previous books. A presidential endorsement followed and the novel started to sell. It has never stopped.

Grahame didn't enjoy his success for long. In 1920, his beloved but troubled Mouse – the inspiration for *The Wind in the Willows* – was found dead on a railway line near Oxford, where he had been a wayward student.

Alistair Grahame had almost certainly taken his own life. And Kenneth Grahame didn't write again.

Grandma Moses

As she reached her late seventies, Anna Mary Robertson Moses had led what was – by the standards of the time – an uneventful life. Born in 1860, one of ten children, she had lived in rural poverty in Virginia and upstate New York as a child and as an adult.

She left her one-room country school at twelve and went to work as a 'hired girl' doing housework on a nearby farm. She eventually met and married a farm labourer and together they had ten children of their own (only five of whom survived).

The family had neither the cause nor the means to go anywhere or do anything else. Her children grew and had more children. Decades passed. Presidents came and went. The world moved on.

And then at the age of seventy-eight, 'Grandma Moses', as she would soon be known, made a fateful decision.

Complaining that arthritis made embroidery too painful, she put down her needles and picked up a paint brush.

Pretty soon she had a small body of work, some of which she displayed in the window of a drugstore in nearby Hoosick Falls, a town of only a few thousand people. In 1939, a New York City water department engineer called Louis Caldor was passing through town. Caldor had a side hustle as an art collector.

He found Moses' paintings in the drugstore and bought them. Then he went round to her house and bought all the rest of her paintings. After that, he set about organising an exhibition of her work in New York.

What did he see in her pictures?

Moses' paintings are simple, colourful portrayals of rural America recalled from childhood. Sotheby's uses words like 'folk art', 'nostalgia' and 'realism'. It is tempting to use words like 'childish', too, but that might well be missing the point.

For a country on the brink of war, Grandma Moses' art offered a reassuring image of back-country America; a quiet place, insulated from national or even international politics. Her pictures were always painted from memory, never from life.

Moses' fame was sudden and complete. She quickly became the best-known artist in the United States. And she continued to work at a furious rate – arthritis or no arthritis – completing around 2,000 paintings in her eighties and nineties.

Her work has been shown in America's great cities and sits in its most important collections and museums. She was compared to Brueghel, whom she'd never heard of, and invited to the White House.

In 1960, Governor Rockefeller proclaimed her 100th birthday 'Grandma Moses Day', saying, 'There is no more renowned artist in our entire country.'

The elderly Grandma Moses seems to have handled her unexpected fame rather well. The *New York Times* described her as 'tiny, lively and mischievous'. She learned to enjoy the attention and relished dispensing old-timey advice, which enhanced both her image as the nation's grandmother and the appeal of her paintings.

'I look back on my life like a good day's work, it was done and I feel satisfied with it,' she said in her book. 'I was happy and contented, I knew nothing better and made the best out of what life offered. And life is what we make it, always has been, always will be.'

Grandma Moses died in 1961 at the age of 101. Her doctor described her cause of death in the kind of folksy manner she would have appreciated: 'She just wore out,' he said.

David Attenborough

David Attenborough has led an extraordinary and, perhaps, an exemplary life. He has been significant for at least three great achievements – but the last of these, and the greatest, has come towards the end of his life.

Attenborough was born in 1926. He took a natural sciences degree at Cambridge and after national service in the Royal Navy and a stint working for a publisher, he found his way to the BBC in 1952.

He did exceptionally well there as a producer, commissioner and executive. He became controller of BBC Two and director of programmes. He commissioned *Civilisation*, *The Ascent of Man* and *Monty Python's Flying Circus*. He led the BBC's adoption of colour television. All of this, then, added up to an impressive career. Enough, certainly, for most.

But for reasons that are easily understood, Attenborough was not content to remain a BBC suit for the rest of his life. So he created and presented *Zoo Quest* – a pioneering natural history programme, which more or less invented the genre – and in time ditched the suit forever to became the most loved broadcaster in Britain.

Attenborough's documentaries have moved hundreds of millions of people around the world and helped audiences everywhere to understand the natural world. This work has always been about far more than the spectacle. Although Attenborough has made some of the most memorable television ever, his programmes have also been serious, groundbreaking and rigorous explorations of the natural world, which have revealed new species and new discoveries.

Another great achievement, then. But in recent years, and in advanced old age (he turned ninety in 2016), Attenborough has re-invented himself as the world's most effective and tireless climate-change campaigner. His unique authority and understanding makes him impossible to ignore. The trust and respect he inspires puts him above any politician.

In 2016, Attenborough met President Obama. 'Part of what I know from watching your programmes,' Obama told him, 'is that these ecosystems are all interconnected.'

In the summer of 2019, Attenborough addressed a British Parliamentary committee, telling the MPs: 'I feel an obligation. The only way you can get up in the morning is to believe that actually we could do something about it. And I suppose I think we can.'

He told the World Economic Forum at Davos: 'I am quite literally from another age. I was born during the Holocene – the name given to the 12,000-year period of climatic stability that allowed humans to settle, farm and create civilisations. We have changed the world so much that scientists say we are now in a new geological age, the Anthropocene – the Age of Humans. When you think about it, there is perhaps no more unsettling thought. The only conditions modern humans have ever known are changing and changing fast.'

Attenborough used the Davos platform to issue a clear warning but also some hope: 'As a species we are expert problem-solvers. But we haven't yet applied ourselves to this problem with the focus it requires.'

Sir David Attenborough was always a conservationist, concerned about the impact of humans on our planet, but he admits he was slow to fully understand the danger we are now in. He is using his remaining energies – and the platform his extraordinary life has given him – to make sure we all catch up with him.

Buena Vista Social Club

In 1996 Nick Gold, a British world-music record producer, and Ry Cooder, an American producer and guitarist, were in Havana, wondering what to do.

Their idea had been to bring together African and Cuban musicians at the EGREM studios – part of Cuba's national record label – and to make an album. But the Africans hadn't shown up. They had run into visa problems.

This kind of thing happens quite a lot in world music, so perhaps Gold and Cooder were used to improvising. They didn't have the Africans, but they did have Cubans – lots of them.

Nick Gold had been working with Juan de Marcos González, a Cuban musician who had been rounding up long-faded stars from Cuban music's golden era – old timers who had enjoyed their first taste of success many decades earlier – to make a record under the name *Afro-Cuban All Stars*.

Juan de Marcos had persuaded the seventy-year-old Ibrahim Ferrer to take part. Ferrer was a *con* and *bolero* singer, who was semi-retired and supplementing his income by shining shoes.

Then there was Rubén González, who first made a name for himself as a pianist in the forties and fifties, but by 1996 no longer even owned a piano – and Pío Levya, who had been making an album in 1953 when the revolution erupted and liked to say you could hear gun shots on the recording.

There were many more. All were immensely accomplished musicians and most of them were very old. Together they had just recorded an album they would call *A Toda Cuba le Gusta*.

Over the following few days – with no Africans in sight – they carried on working; recording another album with an even larger ensemble of legendary Cuban musicians from the distant past.

Now their ranks included Omara Portuondo, who was part of a popular all-female quartet in the fifties, and importantly, Compay Segundo. Cooder said of Segundo: 'He knew the best songs and how to do them, because he'd been doing it since the First World War.'

And so the EGREM sessions yielded a second record. They called this one *Buena Vista Social Club*. And when it was released the following year, something unexpected happened.

It sold. And sold and sold and sold. *Buena Vista Social Club* was a huge hit all over the world. In the end, it sold many millions of copies. No one in world music had ever seen anything like it.

It's hard to say why this happened. It's a wonderful record, for sure, and it captured a once-in-a-lifetime moment. But there had been other great world music records released before it. There have been many released since.

None has ever achieved the kind of success which came to that loose collective of Cuban musicians who, for those few fleeting days in Havana, recorded under the name *Buena Vista Social Club*.

There was a brief tour, which ended in New York City at Carnegie Hall. The film director Wim Wenders used a recording of that performance in a documentary, which prompted even more interest in the group.

All the musicians who took part in the *Buena Vista* sessions had achieved professional success early in their lives. They had all been bright stars of a music scene that was popular in the 1940s and 1950s throughout Cuba and well beyond it.

But the success they encountered in their final years far, far surpassed the fame of their youth. Watch Wim Wenders' film and you get the feeling they were certainly going to enjoy it.

Jessica Tandy

Jessica Tandy achieved success early in her life – and throughout her life – but in old age found herself more in demand, and more famous, than ever. Hers is one of the great acting stories of the twentieth century.

She was born in London in 1909. Her father died when she was twelve and her mother worked long hours at nights to keep the family afloat and, in time, to fund Jessica's place at the Ben Greet Academy of Acting.

It would turn out to be a very good investment; she made her professional debut in 1927 at the age of eighteen and progressed to the West End only two years later.

Although she felt she didn't have the conventional good looks required for leading lady roles – others might well have disagreed – there was no doubting her talent. She played many of the great classic roles and opposite some of the great British actors of her generation, including Gielgud and Olivier.

In 1932 she married the actor Jack Hawkins, whom she later described as 'a wonderful actor, but a rotten husband'. When the marriage failed, and as Britain went to war, she moved to the United States and, eventually, Broadway.

It was on Broadway in 1947 where Tandy gave the performance that, for many, was one of the best New York had ever seen: Blanche DuBois in Tennessee Williams' *A Streetcar Named Desire* (playing opposite a young Marlon Brando).

She wasn't considered a big enough star to take the role in the subsequent Hollywood production, which went instead to Vivien Leigh, who had played DuBois in Olivier's London production of *Streetcar*. But Tandy had made her name in America and would never struggle for roles again.

She had also married again. She met the Canadian-born actor Hume Cronyn in 1942. They would star alongside one another countless times over the subsequent decades, becoming America's great 'theatre couple'.

Tandy's career had spanned sixty-seven years by the time she died in 1994 at the age of eighty-five. She had appeared in over 100 theatre productions, 25 films and many TV dramas. She had won Tony and Emmy awards and as much critical acclaim as almost any other actor. But she had not won an Oscar.

In her final decade, however, she was busier than ever. And in 1989 she appeared alongside Morgan Freeman as Daisy Werthan in *Driving Miss Daisy*. The film was a commercial and critical success, which introduced Tandy to a new generation and brought her global attention.

Tandy's performance was described in *Rolling Stone* as her 'finest two hours onscreen in a film career that goes back to 1932'. At that year's Academy Awards, Tandy finally won an Oscar. She remains the oldest woman ever to be named as 'Best Actress in a Leading Role' by the Academy.

She was nominated again in 1991 for her role in *Fried Green Tomatoes* and continued working till the very end. 'I find that what makes life worth living is the work,' she said, and she meant it.

Raymond Chandler

In 1932, Raymond Chandler lost his job with a Los Angeles oil company. He was 44 years old and had by then bounced through several oil-company jobs, worked in a sports store stringing tennis racquets, and as a reporter. He had also served in the First World War.

He had always written, but never exactly seriously. In 1932 that changed. He fired off stories to pulp magazines and began to refine his craft and find an audience. Finally, in 1939, Chandler published *The Big Sleep*. He was fifty-one.

What's astonishing about Chandler's career is not so much that he came to it late, but the completeness of his success once he had.

He wasn't the only writer of hardboiled detective stories – he might not even have been the best (how about Dashiell Hammett, for one) – but he hit the big-time immediately and never looked back.

His next three books – *Farewell, My Lovely*, *The High*

Window and *Lady in the Lake* – were all hits. Hollywood called. He wrote *Double Indemnity* with Billy Wilder.

Chandler created lasting art within the confines of his genre and influenced any number of other writers. His detective, Philip Marlowe, is one of the most enduring literary creations of the twentieth century.

'I have more love for our superb language than I can possibly express,' he once said. And it showed.

But Chandler was never entirely sure of his place among America's great men of letters. He had a high opinion of his own work – even wondering whether the Nobel Prize might come his way – but a justifiable fear that commercial thrillers would never be taken entirely seriously.

In 1973, the British crime writer Julian Symons wrote that Chandler had come 'to crime fiction late in his career, and was shocked to discover that the people who wrote it were a lesser breed living below the salt, who were expected to bark gratefully for any scraps of notice they received.'

By the time he wrote *The Long Goodbye* in 1953, it had been quite a while since Raymond Chandler had had to bark for scraps of anything.

Chandler's story isn't quite rags to riches – he was, somewhat improbably, educated at Dulwich College in London, a respectable fee-paying school – and it isn't a story of reaching fulfilment after a long struggle either. He doesn't seem to have been particularly happy.

In 1924, the year his mother died, Chandler had married Cissy, a woman eighteen years his senior. (Read into that what you will.) When Cissy died in 1954, Chandler became unmoored.

He attempted suicide, albeit in a notably lacklustre manner, succeeding only in shooting a hole in the bathroom ceiling. He drank a lot and could be wildly obnoxious. (But then, according to some accounts, that had long been the case anyway.)

His biographer Frank MacShane says Chandler 'suffered from bronchitis, strep throats, and skin allergies . . . he also suffered from an allergy that made his fingers split . . . and had his fingers bandaged when he typed.'

Raymond Chandler provides inspiration to anyone approaching fifty who has some residual hope they might write something still worth reading seventy or eighty years into the future. But he may not be a model for how to live comfortably with that achievement once accomplished.

Chandler died in 1959 and is buried in San Diego's Mount Hope Cemetery.

Golda Meir

When Golda Meir became prime minister of Israel in 1969, she was seventy-one years old – and, frankly, she looked it. She was a chain-smoking, deep-lined grandmother, who had lived a hard life. Very little had come easily.

Golda Meir was born Golda Mabovitch in Kiev in 1898. 'I was always a little too cold outside and a little too empty inside,' she remembered. Some days there was only enough food for some of the children. She recalled her sister fainting from hunger.

In 1906, the family emigrated to Milwaukee in the United States. Golda's father found occasional work as a joiner. Her mother sold groceries with Golda's reluctant help.

After rejecting her parents' plans for her to marry a much older man, Golda eventually found a husband, Morris Myerson, another Russian emigre, and Zionism. She persuaded a sceptical Morris to move to Palestine, where the couple were admitted to a kibbutz.

Life on the co-operative farm was backbreaking. She picked almonds, tended chickens and worked herself, she said, to exhaustion. But she impressed her peers and was elected their leader.

This gave her an entree to left-wing politics and the trade union movement, the Histadrut. She moved to Jerusalem and was appointed chief of the Histadrut's political section. After the creation of the state of Israel in 1948, Meir's path to political leadership began to look a little more clear.

She was active in Labor and Zionist politics and in diplomacy, too; quickly becoming an important figure in the political establishment of the new nation. She met Jordan's King Abdullah shortly before the 1948 Arab invasion of Israel, in an unsuccessful attempt to avert war.

Meir would always say that her goal was peace. She wanted Israel to be accepted by its Arab neighbours and to be left alone. 'We say "peace" and the echo comes back from the other side, "war",' she said.

She was appointed Israel's ambassador to the USSR later that year and, in 1949, won a seat in the newly formed Knesset. She was foreign minister from 1956 to 1966, during which time David Ben-Gurion, the prime minister, described her as 'the only man in my Cabinet'.

When Ben-Gurion's successor, Levi Eshkol, died in office in 1969, Meir was elected to replace him. She had been considering retirement. Now her future looked very different. This grandmother would have to lead a young nation through one of the most turbulent periods in its short history.

In October 1973, the combined forces of Syria and Egypt attacked Israel, which was caught off guard – despite its emphatic defeat of Arab forces only six years earlier. The Yom Kippur War was bloody and inconclusive, but it led both Egypt and Israel to take each other more seriously. Egypt's president Anwar Sadat called Meir 'an honest foe'.

The war led indirectly to the 1978 Camp David Accords and Egypt's decision to recognise Israel. But it led to recriminations at home. An inquiry cleared Meir of responsibility for Israel's lack of military readiness, but she chose to bow to the 'will of the people' and step down from power in 1974.

She died four years later, living just long enough to see the Camp David Accords signed.

It later emerged that she had been suffering from cancer for the whole period of her premiership, a fact she had successfully hidden from almost everyone. Golda Meir was an extraordinarily resilient woman.

Stan Lee

When Stan Lee died in Los Angeles in 2018, he was ninety-five years old and at the very heart of popular culture.

Born Stanley Martin Lieber in 1922, Stan Lee grew up in Manhattan. He had aspirations to be a serious novelist, but found work as a junior at Timely Comics, which was owned by a relative.

This turned out to be a well-judged act of nepotism on the relative's part: Timely would become Marvel, and Stan Lee would become its boss.

But he was no suit. He wrote most of the stories himself, deploying pseudonyms to create the impression that Marvel was a larger enterprise than it really was. ('Stan Lee' was one of these.)

'I used to be embarrassed because I was just a comic book writer,' he once said, 'while other people were building bridges or going on to medical careers. And then I realised: entertainment is one of the most important things in people's lives.'

At the time, comics were becoming anodyne. Unhinged Congressmen (every generation has them) were arguing that lurid comics were eroding the morality of the nation's youth.

Batman and Robin (created by Marvel's competitor DC Comics), they said, were clearly homosexuals. Publishers were running scared.

But Lee's characters were never dull. They were – more, perhaps, than any previous comic book characters – three-dimensional: they had superpowers, sure, but they also had flaws. They suffered from anguish and self-doubt. They were socially aware and they could be funny. They squabbled.

That's perhaps why *Iron Man*, *The Hulk*, *Black Panther*, *Spiderman* and all the rest have become some of the most enduring pulp-fiction characters yet invented. You could relate to these guys. Kind of.

In a statement released after his death, Marvel said: 'Stan began building a universe of interlocking continuity, one where fans felt as if they could turn a street corner and run into a superhero.'

Lee himself once told the *LA Times*: 'I wanted the reader to feel we were all friends, that we were sharing some private fun that the outside world wasn't aware of.'

In 1980, Stan Lee moved to Los Angeles to try to push his characters into television and film. He had some success, but nowhere near as much as he had hoped for, or felt his characters deserved.

But in 2000 that changed – emphatically – with the launch of the *X-Men* series and the dizzying number of Marvel releases that followed: *Iron Man*, *Captain America*, *Avengers*, *Thor*, *Black Panther* and many more.

With Stan Lee looking on – by now in advanced old age – the characters he and his comic-book artists had created in Manhattan decades earlier had now become almost the biggest things in popular culture worldwide.

The Marvel film franchise has taken tens of billions of dollars at the box office in the last ten years. It's unclear how much, or how little, of this colossal revenue stream found its way into Stan Lee's pockets before he died.

But it is clear that Stan Lee delighted in being the nonagenerian father of a superhero universe, which continues to have such a firm hold on the imaginations of children (and some adults) everywhere.

Marjory Stoneman Douglas

If you recognise the name Marjory Stoneman Douglas, it's likely to be for the saddest of reasons.

On 14 February 2018, a former student shot and killed seventeen people – fourteen of them pupils – in the Marjory Stoneman Douglas High School in Parkland, Florida. It was one of the deadliest school shootings in US history.

In the aftermath of that shooting, a group of surviving students mounted a campaign for tighter gun laws which was – and remains – as coherent and inspiring as any such campaign before or since.

Born in Minneapolis in 1890, Douglas became a journalist on the *Miami Herald*. It no doubt helped that her father owned the paper; although in those days neither Miami nor the paper were a very big deal.

She went to Europe with the American Red Cross during the First World War, filing stories to the Associated Press, and came back to the *Miami Herald* determined to take a stand against a lot that was wrong with America.

Her editor might have been her father, but she was still pushed towards the diary pages, covering social events and gossip. She responded with fierce pieces about civil rights and welfare. She gave hell to the KKK, the sanitation department, and anyone who thought women should be denied the vote.

In 1917, she led a group of women to address a committee of the Florida legislature. 'It was a big room with men sitting around two walls of it with spittoons between every two or three,' she says in an archive held by Florida International University.

'And we had on our best clothes and we spoke, as we felt, eloquently, about women's suffrage and it was like speaking to blank walls. All they did was spit in the spittoons. They didn't pay any attention to us at all.'

She left the paper but continued to write and to agitate, becoming more and more vexed by the attitude of developers and Florida's politicians to the landscape, particularly the Everglades – which, back then, were seen as worthless.

In 1947, at the age of fifty-seven, she published *Everglades: River of Grass*, an unlikely bestseller that changed the way Americans saw the Everglades forever. Her writing was powerful and her message was stark.

But writing wasn't enough. In 1969, by now seventy-nine years old, she founded the campaign group Friends of the

Everglades and continued to fight for the wetlands' protection.

In its obituary of Stoneman Douglas – she died in 1998 at the age of 108 – the *New York Times* says, simply, that without her the Everglades would no longer exist.

'You have to stand up for some things in the world,' she said. Today's students of the school named for her certainly are. It's a pretty safe bet she would be incredibly proud of them. Her advice to aspiring activists – given in an article thirty-eight years before the Parkland massacre – might have been written for them:

'Speak up. Learn to talk clearly and forcefully in public. Speak simply and not too long at a time, without over-emotion, always from sound preparation and knowledge.

'Be a nuisance where it counts, but don't be a bore at any time. Do your part to inform and stimulate the public to join your action. Be depressed, discouraged and disappointed at failure and the disheartening effects of ignorance, greed, corruption and bad politics – but never give up.'

Mary Wesley

When Mary Wesley died in 2002 at the age of ninety, she was the bestselling author of ten novels with appreciative readers all over the world. Only twenty years earlier, when she turned seventy, she had not yet written any of them.

Mary Wesley was born Mary Aline Mynors Farmar in 1912 to a well-heeled but cold family. Her father, an Army colonel, was largely absent. Her mother placed Mary and her siblings (who, Mary would always maintain, her mother preferred) in the care of a succession of severe governesses.

At one point the fourteen-year-old Mary was apparently left alone by her mother in a Brittany hotel for several months. She received no meaningful education, because it was assumed that she would marry well and wouldn't need it.

In 1937 she was presented at Court, where she caught the docile eye of an eligible peer, one Lord Swinfen, and was soon married – seemingly proving her parents correct in their view that she was destined to be an upper-crust wife.

Swinfen was, by all accounts, a decent enough chap. But the marriage did not succeed, and they divorced in 1945. Her biography explains that the marriage failed in part because of his lack of interest in sex, and her very strong interest in it – an interest she pursued with vigour (with or without the cooperation of Lord Swinfen).

The collapse of the marriage was hastened by the war. Like many other well-bred young women, Mary Wesley was recruited to work for the intelligence services. The war brought despair and privations, but also excitement, meaning and men. In 1944, Mary Wesley went to a party at the Ritz, where she met the journalist Eric Siepmann.

They soon moved in together – scandalising their families – and were married in 1952, after Siepmann's divorce from his first wife. Siepmann was the love of Mary Wesley's life. That much is clear from the intimate letters between the two, which were left to Wesley's biographer Patrick Marnham to publish after her death.

'With you I can become the person I really am,' she wrote. Their years together in a cottage in Devon were her happiest.

Eric died in 1970, leaving Mary both heartbroken and hard-up. By then, she was fifty-eight years old, with very little to fall back on. She took odd jobs to put food on the table and, slowly, started to take writing seriously.

She had written her whole life, even publishing two children's books in the 1960s, but she had never had the confidence to attempt a full adult novel, let alone send one to a publisher. That changed with *Jumping the Queue*, which was published in 1983.

A black comedy of sorts, *Jumping the Queue* tells the story of a suicidal woman after the death of her eccentric husband. She followed it a year later with *The Camomile Lawn*, which became a bestseller, and she kept on going; achieving a level of success in old age that made her not only widely read, but greatly admired.

'A lot of people stop short,' she once said. 'They don't actually die, but they say, "Right, I'm old, and I'm going to retire," and then they dwindle into nothing.' Mary Wesley certainly didn't do that.

Amitabh Bachchan

Amitabh Bachchan's first film bombed at the box office. He persevered for another six years – finding roles in a series of unsuccessful movies – before, on the verge of quitting and heading home to Calcutta, he finally found fame in 1975 with the movie *Sholay*.

To say he 'found fame' is something of an understatement. Over the next ten years, Bachchan made a succession of films that turned him into the biggest star India had ever seen. He played 'angry young man' characters; working-class heroes who could stick it to the man.

The films resonated in an India roiled by 'The Emergency' and the widespread abuse of government power. Bachchan met the ravenous demand for his movies by turning out as many as humanly possible – once, it's said, filming twelve productions simultaneously, hopping from studio to studio and scene to scene without let-up.

Bachchan's films were exactly the kinds of fantasies many Indians needed. Amitabh Bachchan himself was the perfect hero: sensitive but tough; handsome but relatable; and famously modest about his success. Indians rewarded Bachchan with unreserved adulation.

In 1982 he was nearly killed in an accident on set and his slow recovery was treated like the aftermath of a natural disaster: temples, mosques and churches all over the country prayed for him. The prime minister sat at his bedside.

Bachchan was thirty-three years old in 1975 when he made his breakthrough. So although he had paid his dues longer than many leading men in Bollywood, he wasn't exactly in late life when he found success. The reason he qualifies for inclusion here is that he would have to do it all over again, much later.

By the mid-1980s, Bachchan's appeal was beginning to wane. He was looking a little old to be an angry young man. Other actors were stealing India's affection. He chose some questionable movies. And then he made two big mistakes.

He was persuaded to enter politics by the Gandhi family and in 1984, a little reluctantly, he became a member of parliament. Bachchan had been the biggest movie star in Asia. He wasn't ready for the grubby theatre of Indian party politics.

'I ended up in Parliament,' he said, 'and soon discovered that emotion really doesn't have any place in politics. It's a much more intricate and complicated game, and I just didn't know how to play it.' He resigned in 1987. But his brand had been dented.

Then, a few years later, he went into business, setting up the Amitabh Bachchan Corporation – a production company. Business, it seemed, wasn't for him either. The company failed.

By the end of the millennium, Bachchan was nearly sixty years old and in trouble. 'It was essential for me to do something,' he later told the *Independent*. 'I was facing bankruptcy, court cases, creditors, a failed company and a failed career.'

Salvation came in the surprising shape of *Who Wants to Be a Millionaire?* – the British TV format that was launched in India in 2000 under the name *Kaun Banega Crorepati*. Bachchan didn't win it. He agreed to present it.

It might have seemed as if this once great actor, who had appeared to carry the entire Indian film industry for a time, had been humiliated. A game-show host? But when the series launched, things started to look rather different.

It was a huge hit – a sensation – which led to Bachchan becoming the highest-paid star in Indian TV history, and, more importantly, to the energetic resumption of his acting career.

He did not pick up where he had left off. His failure and resurrection had somehow enhanced Bachchan. He was no longer an angry young man; he was a wise and immensely charismatic older one.

He has made a number of films since 2000 that are every bit as good as any of his earlier work, and some that are far better. Amitabh Bachchan, now well into his seventies, is more beloved by Indian audiences than ever.

It's difficult for Hollywood audiences to understand Bachchan's status in Asian cinema, because no Western actor occupies such a rarefied position. But Bachchan, throughout his first taste of success – and even more emphatically throughout his second – has retained a modesty and indifference to age that are a part of his appeal.

'Basically, I am just another actor who loves his work,' he says, 'and this thing about age only exists in the media.'

Barbara Hosking

In 1947, at the age of twenty-one, Barbara Hosking left what she's described as a 'miserable' childhood in Cornwall and moved to London. Her plan was to pursue a career in journalism.

Instead she found her way to the Labour Party press office where, apart from a few years spent – somewhat unexpectedly – working for a mining operation in Tanzania, she became thoroughly immersed in Labour politics.

After serving for a time as a Labour councillor in Islington, the natural next step for this bright, capable and now politically connected woman was to run for a Westminster seat.

She was considered for the Stroud constituency. But the interview process left her wondering whether she was really cut out for party politics at all.

The interview focused on her support for the Campaign for Nuclear Disarmament – hardly an unfashionable position in

Labour circles – and whether she would, if required, set aside her personal political principles if they were ever to conflict with the party line.

Hosking understood the need for party discipline but she also, in that moment, understood that it wasn't for her. She had no intention of leaving her political convictions at the door. She was offered the seat – the panel apparently impressed by her combative performance – but declined it.

'I now was aware for the first time in my life that I thoroughly disliked political compromise,' she wrote in her 2017 memoir, *Exceeding My Brief: Memoirs of a Disobedient Civil Servant*.

She went instead to a part of the political system that requires not blind loyalty to the whips' office but neutrality: the civil service. She quickly found herself in the Downing Street press set-up, working closely with Harold Wilson and then Edward Heath.

Her proximity to power gave a front-row view of history. She was with Heath when he signed the Treaty of Rome in 1973 (an achievement she has watched, with a heavy heart, more recent Conservative prime ministers dismantle).

But she also found herself with a very clear view of Britain's class injustices. Her modest origins and faint Cornish accent made her a rare outsider in Whitehall, where polished Oxbridge graduates – almost all men – dominated.

'You can tell a person's class or their background by their haircut, by their shoes, they don't have to open their mouths,' she told the *New Statesman* in 2017. 'It's a terrible thing, in that it's endemic in this country.'

Hosking went on to important roles in broadcasting and continued to call out sexism and unfairness wherever she found it.

At the Independent Broadcasting Authority in the late 1970s, she was furious to discover that her male deputy was being paid more than her – a situation she rectified within hours after marching into the director-general's office to make her view of the situation clear.

Hosking had climbed into the British establishment despite having none of the obvious advantages: she was from Cornwall, not the South-East, she had a modest education, no family connections and she was a woman.

And there was a further obstacle, which – in the Britain of the 1960s and 1970s – could have set her even further back: she was gay.

Barbara Hosking didn't come out until 2017, when she was ninety-one years old. She did so because she felt she couldn't publish an honest memoir without making her sexuality clear. It seems not to have been a big decision for her. But it was seen as an important moment, nonetheless.

She was surprised to discover both that some of her close friends had no idea (she had thought it was rather obvious) and that her decision to come out was treated as something to celebrate – reflecting, perhaps, how far attitudes to sexuality have changed in Britain over the course of this one woman's long life.

Charles R. Flint

Charles Ranlett Flint was born in Maine in 1850 to a family of shipbuilders and operators. Some sixty-one years later, he created IBM. There were a number of steps in between.

In truth, Flint was hardly a late-achiever. His entire life was extraordinary – almost improbably so. But had he retired at sixty, it's quite likely his many successes would by now be long-forgotten. IBM ensured his legacy.

By the time of his death in 1934, IBM was well on its way to becoming one of the largest and most enduring enterprises in US corporate history.

Flint worked with his family in shipping for a time, but soon found he could make more money brokering deals between and on behalf of other corporations, particularly if those corporations were in the business of supplying arms.

He funnelled munitions to the Russians and the Peruvians and sent the Wright Brothers aircraft to the Germans. He sold ships to the Japanese for use in their war against China and then sold ships to the Russians for its war against the Japanese.

Along the way he dealt in guano and nitrates, set a world water-speed record in a steam-powered vessel, learned to fly at a time when doing so was purely experimental and – crucially – began to develop a view of the corporate world that we would regard today as obvious but which, at the time, wasn't.

Flint's idea was that companies should be as big as possible. He was a pioneer of mergers and acquisitions. He saw an argument not only for economies of scale, but for the value of intangible assets like market share and goodwill which, when brought together in powerful corporations, could be efficiently exploited.

He proved his point several times, consolidating industries from rubber – financing the creation of the United States Rubber Company – to caramel production. He didn't always get it right.

His attempts to consolidate America's competing electric-light companies failed. But, for the most part, it seems Flint had a high old time becoming, of course, immensely rich and influential in the process.

He also believed in what would come to be called 'shareholder capitalism', arguing that large firms should be owned not by a handful of the exceptionally rich, but by a wide pool of much less well-heeled investors through the US stock market.

Flint was, in short, both a brilliant dealmaker and a prescient observer of American business.

By the early 1900s, Flint had interests in a vast portfolio of companies, including International Time Recording, which made time clocks, and Computing Scale Company of America, which provided weighing and pricing equipment (and cheese slicers) to store clerks.

He merged these two companies, explaining that – cheese slicers aside – they were clearly related. And then he added another: the Tabulating Machine Company, which was facing an uncertain future in the face of competition.

In 1911, Flint re-branded the new conglomerate Computing-Tabulating-Recording (CTR) and in 1914 made an inspired appointment, offering the leadership of his fragile new enterprise to Thomas J. Watson, who worked for a rival firm called National Cash Register.

CTR would be re-branded International Business Machines Corporation (IBM) in 1924. Watson is rightly credited with turning IBM into the world-dominating organisation it quickly became. Others have kept it there, despite the rapidly changing technological world in which it operates.

But Charles Ranlett Flint was the sixty-one-year-old arms trader, yacht racer, pilot and dealmaker who made it possible.

Shigeo Tokuda

In 1994, Shigeo Tokuda retired after a solid but unspectacular career as a travel agent. He was sixty years old, happily married, a father, and looking forward to spending more time watching pornographic videos.

Reluctant to buy the videos from local shops, however, Tokuda began to visit porn studios to pick up boxes of VHS tapes himself. There he befriended a porn director, who made the somewhat surprising suggestion that perhaps Tokuda should not only watch pornographic films, but appear in one.

Tokuda had to think about it for quite a while. After all, what would his wife say? But the director was onto something. Japanese society is ageing: there are a lot of older people, like Tokuda, with time on their hands.

Within ten years, Tokuda had starred in literally hundreds of porno films, alongside co-stars aged from eighteen to seventy-two.

The plots of these films are unconvincing: Tokuda as a company boss who finds young female employees happy to do almost anything to please him; Tokuda as an elderly gentleman reliant on his attractive young carer. One thing soon leads to another.

Now in his mid-eighties, Tokuda is still going, providing inspiration to a generation of older Japanese men. He attributes his enduring virility to hiking and a diet of vegetables and eggs. He is known in Japan as 'The Legend Grandpa'.

Films depicting weak old men with obliging young women have become a hugely popular genre in Japanese porn (they account for about a third of the whole porn market there) and today Tokuda stands, a little unsteadily, astride the genre like a naked and wrinkled colossus.

He is philosophical about his late success. 'I first learned about sex in the postwar period, when we didn't have anything,' he told *Vice*.

'Everybody was so focused on living, studying, graduating, and working that neither my classmates nor myself were all that interested in sex. So for my generation, it's all about getting back those lost dreams of our youth.'

Tokuda, which isn't his real name, of course, tried for a time to keep his new role a secret from his family, telling them he was finding work as an extra in conventional TV and video productions.

But once he became 'The Legend Grandpa', the secrecy became much harder to sustain. One day, his daughter found a fax offering him a scene, setting out the plot's requirements in explicit detail. She was, he told the *Guardian*, 'speechless for a while'. Tokuda's wife, he says, would simply prefer not to know too much about it.

'She occasionally suggests it's about time I quit, but we have never argued about it. As long as I don't bring up the subject and treat it more like a hobby, I can get away with it.'

Judith Kerr

In 1933, the journalist Alfred Kempner picked up a rumour that his arrest was imminent. He lived in Berlin. He was Jewish. And he was a vocal critic of the Nazi Party.

He fled with his family to Switzerland, France and, finally, in 1936, to London – where, after leaving his home and income behind, he was forced to rely for a time on the kindness of strangers.

Kindness is a thread that seems to have been woven into the entire life of Alfred's daughter, Judith Kerr, who was thirteen in 1936 when the family started its new life in London.

She spoke often of the kindness the immigrant family encountered (despite their German accents). Later she would delight generations of British children with her clever, gentle – and kind – stories.

But in 1936 that was still a long way off.

First she had to: finish her itinerant education; endure the Blitz; become a script editor at the BBC; meet her future husband in the canteen; buy a house; produce two children; and raise them (rather well, judging from their subsequent successes) to school age.

Eventually, in middle age, Judith Kerr sat down to write and illustrate in earnest. In 1968, at the age of forty-five, she published her first book, *The Tiger Who Came to Tea*, telling the story of a tiger that interrupts a little girl and her mother's otherwise uneventful afternoon.

Did the Tiger – and his unexpected knock on the door – represent the Gestapo?

No, Kerr would say, he represented simply a tiger who came to tea.

Kerr's uncomplicated drawings of the ordinary home in which the Tiger ate all the food and drank all the drink were faithful renderings of Kerr's own house in south-west London. Somehow children knew that Judith Kerr's home, real or imagined, was a good place to be.

Then came *Mog*, a forgetful cat who unknowingly saves her family from a burglar, inadvertently pushes a baby out of the path of a car, and bites the vet. These and other stories sold 10 million copies, making Judith Kerr one of Britain's most successful authors.

Before her death in 2019 at the age of ninety-five, Judith

Kerr began to worry that a creeping unkindness was emerging in Britain.

She was shocked by evidence of antisemitism in public life. She couldn't understand why Britain would turn its back on a European project that had kept peace on a Continent once ablaze with war – one she clearly recalled.

She told *Country Life* magazine: 'Those of us who survived the Second World War have lived through probably the safest and most affluent period in history. But, because I'm so old, I remember how quickly it can change. I remember the war and how, suddenly, the world was a totally different place. I walk about and look at people, out with their children and walking about and shopping, and I think: do they realise how fragile it all is?'

Ronald Reagan

Capturing the life of a president in around 500 words is a hopeless task.

But for our purposes, the point about Reagan is not so much that he became the 40th President of the United States of America, but that he did so – for the second time – at the age of seventy-three, making him the oldest American ever to win a presidential election. (So far.)

Ronald Reagan was born in 1911 in Illinois, by all accounts an excellent example of young American manhood: tanned and lithe, wholesome and likeable, a football player, lifeguard and high-school actor.

After a short career as a sports radio announcer – that voice! – he went for a Hollywood screen test and was signed to a studio in 1937. He would appear in over fifty films. They were not all great films. But they made him a recognisable figure across the United States.

Reagan came to politics through his presidency of the Screen Actors Guild. Union work on behalf of Hollywood actors isn't exactly the sharp end of politics; an unlikely platform, on the face of it, from which to launch a national political career.

But the job gave him exposure as a speaker – and, boy, could he speak. In a 2004 *New Yorker* piece, Edmund Morris – who became Reagan's official biographer – wrote: 'Merely by breathing, "My fellow-Americans", he made his listener trust him.'

Union work also led Reagan to encounter more conservative political views, which he quickly adopted. He was, in truth, a far better political speaker than actor. His folksy charm and rare charisma (as those Hollywood scouts had detected back in 1937, the camera loved him) propelled him to the California governor's mansion in 1966 and 1970 and, eventually, to the White House in 1981. He was seventy years old.

Reagan served two terms. Plenty happened. The Reagan economy boomed. He was shot. His policy of 'Peace Through Strength' arguably helped steer the Cold War towards a thaw – and on the side he meddled in support of anti-communist movements in several countries.

By the time he left office, Ronald Reagan was nearly eighty. Age had been an electoral issue throughout his presidency. But while in 1981 he had appeared old, he was the right kind of old: he looked genial, avuncular and steady. He took doubts about his age and flipped them.

During a 1984 campaign debate against the fifty-six-year-old Walter Mondale, he neutralised the issue with one devastatingly brilliant line: 'I want you to know that . . . I will not make age an issue of this campaign. I am not going to exploit, for political purposes, my opponent's youth and inexperience.'

By 1989, there were rumours that he lost focus in briefings, that he was no longer fit for office – perhaps, even, that he was in the early stages of dementia.

In his *New Yorker* piece Edmund Morris, who met Reagan regularly during his second term, tried to set the record straight:

'I never saw any sign of cognitive dementia. There were, to be sure, days late in his Presidency when he drifted off, as old men do. On 29 May 1988, for example, he emerged from an extended one-on-one with Gorbachev unable to recall a word that had been said. But such lapses were rare, and could usually be ascribed to fatigue.'

In 1994, Ronald Reagan issued a letter to the American people in which he announced that he was, in fact, 'one of the millions of Americans who will be afflicted with Alzheimer's disease.'

The letter was – typically – direct, elegant and even optimistic.

'I now begin the journey that will lead me into the sunset of my life. I know that for America there will always be a bright dawn ahead.'

Barbara Woodhouse

Barbara Woodhouse must rank high in the list of all-time most unlikely celebrities. And it all happened right at the end of her life.

At the start of 1980, she was seventy years old and more or less an unknown. Just a few months later she had become – and would remain until her death eight years later – one of the most recognisable women in the world.

How did she achieve this rapid ascent to fame? By commanding dogs to 'Sit!' on television.

Woodhouse was born Barbara Blackburn in 1910, the daughter of an Irish clergyman. When her father died, the family moved to England and Barbara's mother boarded dogs to help make ends meet. Woodhouse would say she learned to quieten the dogs simply by talking to them.

The next fifty years seem to have passed in unspectacular fashion. She got married and divorced and married again. She worked for a time in Argentina. She wrote books with titles such as *The A to Z of Dogs and Puppies*.

And then the BBC put her on television. A lot of people will have watched her ten-part BBC Two series, *Training Dogs the Woodhouse Way*, simply because, in the Britain of 1980, there was virtually nothing else to do. Woodhouse herself was not an obvious fit for the medium.

She had the demeanour of a not-entirely-lovable maiden aunt, who much preferred canine company to that of humans. Her harshest commands were generally reserved for the dog-owners, who always looked more uncomfortable than their pets. Her catchphrases – 'Sit!' and 'Walkies!' – seem, in retrospect, a little uninspired.

So what happened next is hard to explain.

In short: the series was a phenomenal success in the UK and the US, making Woodhouse an instant celebrity of international proportions. Before the year was out, she was named Britain's TV personality of 1980. In 1982, she published the bestseller *No Bad Dogs: The Woodhouse Way*.

She believed a dog's poor behaviour could usually be explained by weak owners failing to make their pet understand its low place in the family hierarchy. Any dog could be trained with the right approach, she said.

She described her method as relying on a kind of telepathy. 'These things are not always born in people,' she explained. 'They can be developed as any sense or gift can be developed.'

The *New York Times* reported that her way of getting to know an animal was to breathe into its nose.

It's not entirely clear what, other than instinct, qualified Woodhouse to become the world's most influential dog trainer. Some research has cast doubt on the efficacy of her methods. But her techniques were widely imitated.

Whether or not she really understood the minds of dogs, she certainly came to understand the power of television.

Anyone who met the ageing and headmistress-like Woodhouse before she found TV stardom could surely never have guessed where her dogs were about to lead her.

Harry Read, John Hutton and Tom Rice

In the early hours of 6 June 1944, Harry Read crouched at the open door of an aircraft as it flew in darkness over enemy-occupied France. He was a twenty-one-year-old wireless operator with the Parachute Regiment, and this was D-Day.

In his final briefing, Read had been told the Paras were expected to lose half their men on landing. He told the *Guardian* he sat alone, contemplating this news, and came to the conclusion that he would 'do everything I could to live up to being a Para in enemy country. I wouldn't surrender.'

When Read hit the ground, he was pulled by the heavy wireless battery strapped to his leg into a deep, flooded trench. He almost drowned before firing a single shot. Many others did drown. The German infantry had deliberately flooded the landing zone.

Stirling-born John 'Jock' Hutton signed up for service the first chance he got. He was nineteen years old on D-Day, serving with the 13th Lancashire Parachute Battalion.

He made his jump over Pegasus Bridge and was soon engaged in heavy fighting. He told the *Herald*: 'Three weeks after the landings, I was on a night patrol. Germans saw us in the moonlight and threw grenades. Such was the noise, I didn't realise I had been wounded. It was only when I got back to my hole in the ground, I couldn't feel my legs.' Shrapnel from that attack is still lodged somewhere in his stomach.

Tom Rice was serving with the US Army's 101st Airborne Division in 1944. His jump over Carentan was, he says, one of his worst.

'I got my left armpit caught in the lower left-hand corner of the door,' he told the *Associated Press*. 'So I swung out, came back and hit the side of the aircraft, swung out again and came back, and I just tried to straighten my arm out and I got free.'

None of these men knew each other. They were three of some 18,000 Allied airborne troops dropped into enemy territory shortly after midnight on 6 June 1944. Their difficult job was to prepare the way for the seaborne assault, which would follow later that day.

This would be the largest seaborne invasion ever mounted – with 7,000 vessels landing 132,000 troops (4,500 of them killed on day one) – and it was a decisive turning point in the Second World War.

On 6 June 2019 – seventy-five years after D-Day – all three men parachuted into France again. This time Harry Read was ninety-five, John Hutton was ninety-four and Tom Rice was ninety-seven.

They jumped close to their original landing sites and were watched by other veterans, families and even world leaders as they parachuted into events to commemorate the D-Day landings.

Tom Rice told the *Associated Press*, 'I feel great. I'd go up and do it all again.' John Hutton said, 'It's great to be back on French soil.'

Harry Read, a great-great-grandfather and retired Salvation Army officer, said: 'I thought the jump was brilliant. The jump was wonderful in every way. I feel good. My health is good, and my mind is still ticking away.'

These three men had what it took as young men in 1944, playing their part in a terrifying mission that would leave many of their brothers-in-arms dead. They each proved in their nineties that, although they may be frailer, they are no less courageous now.

Index of Names